CW01501579

LEAN SIX SIGMA

3 BOOKS IN 1 - THE ULTIMATE BEGINNER'S,
INTERMEDIATE & ADVANCED GUIDE TO LEARN
LEAN SIX SIGMA STEP BY STEP

JAMES TURNER

CONTENTS

LEAN SIX SIGMA

LEAN SIX SIGMA

LEAN SIX SIGMA

LEAN SIX SIGMA

THE ULTIMATE BEGINNER'S GUIDE TO LEARN LEAN SIX SIGMA STEP BY STEP

INTRODUCTION

All businesses strive to ensure they remain profitable, especially in the face of competition and unfavorable business conditions. Over the years, businesses have invested a lot of resources into quality management to make sure that they remain competitive and control a fair share of the markets.

Many quality management protocols have been advanced over the years. One of these is Lean Six Sigma. This is an approach that combines two quality management methodologies, Lean and Six Sigma. Lean Six Sigma, therefore, is a hybrid.

The focus of Lean Six Sigma is to reduce variations present in the business processes and, at the same time, eliminate waste and ensure customers are satisfied. Lean Six Sigma focuses on value addition. Any process, service, or product that adds no value to the production process or that does not contribute to customer satisfaction is deemed surplus to requirement.

Lean Six Sigma is structured as an efficient method of solving problems. Lean Six Sigma proposes two frameworks for problem solution and value addition, DMAIC, and DMADV. However, DMAIC is strongly aligned with Lean Six Sigma, given the emphasis on waste elimination.

Lean Six Sigma is strongly appreciated in the quality management realm for a lot of reasons. One of these is the fact that it uses scientific methods to address issues. It is a fact-based system that emphasizes the need to prevent defects in the first place instead of detecting them.

The problem with defects is that the company has to allocate more funds toward correcting the problems. This ends up tying up operational capital. Defective products that make it to the market also have the risk of customer dissatisfaction. If customers are not happy with what is presented to them in the market, they feel the company does not take them seriously. In such a case, it is within their right to choose a different company to work with.

While most people are just learning about Lean Six Sigma, it is not a new concept. Lean Six Sigma has been around for years but has evolved over time, just as business needs and strategic goals do. We live in a business world that is dynamic and that keeps changing over time. One thing that changes more than anything else is customer needs. Customer needs change in relation to the customer's environment. Most, if not everything, within the customer's environment cannot be influenced by the company. However, these factors influence the decisions of the customer and their interaction with the company. The company, therefore, has to do all it can to stay relevant and meet the customer needs.

Lean Six Sigma incorporates waste reduction and reducing process variation. These are two crucial factors that have a direct influence on customer requirements. When combined together with the organization's strategic plan, you have a process that can help the business achieve its full potential. This process helps management optimize how resources are used, reduce unnecessary costs, and improve their returns.

In this book, we will cover Lean Six Sigma in-depth, giving you the necessary tools and information you need to go ahead and begin your lessons in quality management. The information herein should guide you and help you succeed and become a highly sought-after expert at Lean Six Sigma.

1

BRIEF HISTORY OF LEAN SIX SIGMA

Quality management has evolved over the years. From basic inspection, quality management can now be determined by modern interpretation through Lean Six Sigma. This is important, especially when determining the tools and processes that are necessary to meet the fundamental improvements in your business.

Most of the quality controls in production industries can be traced back to the 1940s in Japan (Allen, 2018). Supervisors would have to inspect an employee's work and decide whether to reject or accept it. Over time, the production industry expanded, and the decision-making process became more rigorous. This necessitated the need for an all-around inspection role.

Another challenge arose with the creation of inspection roles. Initially, supervisors only needed to accept a simple product and approve it as ready for the market. However, as production expanded, the need to prevent defects in manufacturing that would end up in product rejection was imminent. This is how quality management came to be.

During the World Wars, the industrial capacity in Japan had been decimated. The Japanese manufacturing industry was filled with illiterate workers, and there was a reputation for imitative products that were very cheap. To change this situation, the people involved made an industry-wide call to invest in quality management.

After the war, the quality management practices that were employed in Japan became popular. By the 1960s, quality management was one of the leading job descriptions. Japan had eliminated the flaws in their production system and was producing better quality output than was present before and during the wars.

Apart from producing superior products compared to the competition in the West, Japan also enjoyed low prices. Europe and the US were attracted to Japanese products, and the export market grew from this. Japan was reaping profits from quality management, and the Western market could not do without her products.

It was not until 1969 that Europe, Japan, and the US championed the first international conference on quality management. During the conference, quality management factors that affect business performance were discussed. These include planning, taking responsibility by management, and organization by Feigenbaum.

Quality management became a hit in the West, given the desire to meet and beat the quality that was coming from Japan. By the 1980s, almost all production centers in the West had employed total quality management in different sectors. The benefits were amazing.

Today, quality management involves the performance of the organization in general and the important role in the performance of the company. Quality management has been studied in-depth, and recommendations have been proposed to address specific aspects of the production process. Organizations that have implemented these procedures effectively have enjoyed sheer success over the years, and they continue to bask in their glory. One of the quality management models is Lean Six Sigma.

2

PROCESS MINING AND THE CORE PRINCIPLES OF LEAN SIX SIGMA

When building your business or career, Lean Six Sigma will be useful as a means of improving quality management in your endeavors. You want to build a career or business that is supported by strategic principles, which will help you succeed in the future. In business, you will encounter a lot of variables to measure and factors to consider. How do you tell which one is better than the other or where to begin? How do you determine who should shoulder responsibility for different improvement processes?

For lack of understanding, a lot of people and even businesses go into important decision making processes blind. You should instead consider Lean Six Sigma. This is an approach that helps you find a balance between shoring up processes and getting the right focus to help you achieve your goals. In the long run, you can be happy with reduced defects from processes and procedures, which will save you time, money, and other resource wastage.

Principles of Lean Six Sigma

Each process has guiding principles that make it succeed in what it is applied to do. In Lean Six Sigma, the following are the core principles that will guide implementation:

Prioritize Customer Needs

When you work with any Lean Six Sigma approach, it is vital that you step back and reflect on what you are working toward. What are your reasons for doing what you are working on? To improve most processes, you must understand the scope of the process. In Lean Six Sigma, you are not just going to work around a single process but combine different processes to move your business ahead.

While you have to improve each segment in the business on its own, look at the collective gains that you will make from improving all components of the business that are linked together. The most important thing in the summation of all the components of your business is that your customer must derive value from the business they conduct with you. Other than that, you should steer the business in a position where it can offer the value the customers need.

Look at it this way—why would you invest so much into optimizing a process and improving products and services that no one is interested in? Quality constitution in the business should revolve around improving processes and products. Taking that into consideration, you should also appreciate the value of your products and how you can improve the same.

Businesses thrive in the face of internal and environmental challenges. Customer value is an external force that you have zero control over but will play an essential role in determining your success. It can also influence your activities and how you go about them.

Another way of looking at it is in terms of waste management. Waste reduction is an integral part of Lean Six Sigma. When you are planning to identify and eliminate waste in your processes, you are looking at things that do not add value. This is why it is easy for you to understand what is useful to your processes. You know what your value is and how to attain it.

Without a proper assessment of value, you cannot conclusively tell what is a waste to your processes; hence, you will struggle to make any changes that will give a positive impact on your business. Start by understanding your customers, what they need, and how you can offer what they need. Remember that improving processes in the workplace is not an end but a means to an end.

Introduce Fluidity in Processes

At the core of Lean Six Sigma is process management. This approach takes away a chance approach or gambling with possibilities and introduces functionality in management. Processes must smoothly run for you to achieve your desired goals. As a Lean Six Sigma professional, how do you make sure this happens?

First, you must identify and understand the problems. Before you begin, take time and look for any bottlenecks in the processes. Bottlenecks are evident in a situation where you have one segment of the entire process underperforming, to the point where all processes that are hinged to it suffer the same fate.

Bottlenecks interfere with process flow in any organization. You can build on the gains in process mining to identify processes that have bottlenecks. Once these are identified, you can address the challenges they raise and how to mitigate these in the future. For a small or medium-sized entity, you can address the identified bottlenecks by looking at the root cause.

Value Addition through Waste Reduction

When you know your value, nothing can stop you from getting what you deserve. The same applies to Lean Six Sigma. To reduce waste, you must know what you are worth or what you are working toward. From there, you get rid of anything that does not contribute toward the goals you are working at.

In most production or manufacturing plants, the common wastes are overproduction, time wastage, logistical challenges in transportation, stock taking, defective products, processing, and movement waste. Each waste is unique, and you must address it in a different approach. However, in some cases, you can use the same approach to deal with different kinds of wastage.

When handling waste, you must also know whether you have succeeded in eliminating it or not. This means you should start by making waste tangible. Once waste is tangible, you become aware of its presence and push for accountability. Someone has to be responsible and held accountable for that waste. Accountability is important because whoever is in charge will be sticking out like a sore thumb, and no one likes that. Quantify the identified waste, and find a way to either reduce it or get rid of it altogether.

Embrace a system where your employees try to avoid waste altogether, instead of spending to get rid of it. You can employ systems that will reduce that waste instead of having to invest in eliminating it.

Eliminate Variation

In as much as there is always a reason to champion differentiation, it brings about variation, which in essence introduces redundancies and waste in the production process. It is advisable that you consider standardization, given that it has been proven in the past to improve morale, productivity, and quality in many processes.

Looking at Lean Six Sigma, you will find that it is easy to see why the elimination of variation is a priority. Lean Six Sigma is about improving processes. Variation creates a lot of waste. Standardizing processes allows room for optimization, which makes it easier for you to get things right. Standardization is useful as it helps you reduce the level of defects in the processes.

You can introduce checklists to help standardize different functions in the business processes. Through checklists, you will also make some activities mandatory. The good thing with checklists is that even the most mundane of checklists will still give you a record of work that is completed, and they can be implemented in any organization.

To eliminate variation and optimize the processes in your business, you can also consider the following:

- Introduce procedure that mandates people to own up to responsibilities.
- Create assignments that encourage collaboration, where multiple people are brought together into a process.
- Use conditional logic to account for complexities in certain processes and variations that result from the same.
- Use template overviews to see all the checklists which have been implemented.

Collaborate

Collaborative approaches are necessary if your company is to succeed, especially in an instance where teamwork is needed. Analysts and managers can easily make decisions on processes depending on their unique perspective. More often, their view is from a managerial perspective. However, for the decisions to be effective, it is wise to include employees who will be a part of the process to get involved. This creates a sense of inclusivity, and the employees get a different insight into the project.

Workers who are a part of the process each day know a lot about the process. Therefore, their input in the decisions concerning the process will go a long way in improving the process than locking them out of decision-making. Besides, in Lean Six Sigma, there are different levels of membership, and each person is held accountable for different roles. You should try and get your team into the project as active participants, especially in consultation and discussions around the process problem. By building a positive rapport with the people involved in the processes, you stand a better chance of succeeding with the project.

Exert Scientific and Systematic Efforts

Lean Six Sigma is popular for one main reason—it helps you create a system through which you can improve processes in your workplace. Lean Six Sigma is a data-driven process. All decisions made are leveraged on data analysis, which brings a scientific approach to business processes. The DMAIC process that is the pillar of Lean Six Sigma addresses solutions to problems by calculating the possibility of success and quantifying your efforts to succeed.

Understand What You Do

Before you start improving processes, you must first understand what processes are involved and how they run.

Assuming you do not know what process is running, you cannot fix what you do not understand.

Processes and workflows in any organization must be properly documented. In the documentation, you should espouse the ethos of your business and management process. You can consider process mining for this. Process mining helped a lot of professionals in the past understand what their companies were working with.

But what is process mining? Process mining is a situation where you incorporate analytical software to study and understand the work you do and the team involved so that you know the order in which tasks are performed (Aalst, 2016). Most of the process mining software can be used to map event logs, which will give you a deeper analysis of the activities, and from here, you can document them accordingly.

With the processes properly documented, you have the tools at your disposal to find out the work that is done and whether it is done accordingly. This is to check conformance. Conformance checking is necessary if you are to fully understand what the business processes are. It will beat the purpose of process mining if you do not follow up to make sure things are done accordingly.

Process Mining

Business is about processes and making sure that they work toward achieving your ultimate goal. You must find the best way to analyze, build, and track all processes to stay ahead of the game. There are many ways of managing processes which can be implemented, and one of which is **process mining**.

Process mining is a data-driven approach that helps you get a deeper insight into your projects. Through process mining, you have a better shot at generating, interpreting, and optimizing the business processes in your organization. This is a process where you use specific software to get a closer look at what your employees do at work and have the software describe the employee activities for you as processes.

In process mining, you will be using event logs to see what your employees are up to. You can discern the procedures that they take before a task is completed, processes that are introduced automatically into your system, and also identify the stages where bottlenecks are introduced into your business process.

How to Implement Process Mining

Process mining can be implemented at work in one of three ways:

- Performance analysis
- Checking process conformance
- Automated business process discovery (ABPD)

Automated Business Process Discovery (ABPD)

ABPD is a subprocess of process mining, though, in essence, it is a general term that can be used to refer to process mining as a whole. In ABPD, you will track event logs to find out what tasks your employees are working on and define all the processes that are involved in the said tasks.

While programs like these make your work easier, learning about what your employees are doing should not be your end goal. The ultimate goal should be the changes that you can make at work with the knowledge that you have.

ABPD is useful for companies that are yet to come up with a clear operating protocol. If your company has never defined operating procedures, this would be a good place for you to start. It would be useful because it takes away the redundancy of having to wait on several stakeholders to approve procedures that they might not understand.

The process of mapping activities is a straightforward and nondisruptive one, and you can identify processes without worrying about committing a lot of resources to the task. In case you have plans to advance your business into business process management or if you are planning to get ISO certified, process mining can save you a lot of time and effort.

Checking Process Conformance

Conformance checking follows process definition. You have already figured out the processes that take place in your organization. The next step is to make sure that everyone involved keenly follows their processes, and you must also make sure the processes are not redundant.

Conformance checking is about evaluation. You will monitor the event logs to ensure that all the actions are in accordance with the process model you implement in your workplace. If, for example, you have realized that to complete a given process, employees on the task must commit to steps A, B, D, and E. However, you notice from the logs that your team is performing step C, which is not necessary. You can let them know and make the necessary changes.

Therefore, while ABPD helps you identify processes within the organization, conformance checking allows you to ensure your employees are keenly following the processes outlined. Conformance checking will only work for a company that already has its processes and procedures in place.

Ultimately, conformance checking and ABPD go hand in hand, because they complement one another. They both keep event logs, and from these logs, you can use the information gained to compare what you had planned for with the true picture on the ground.

Performance Analysis

Performance analysis combines ABPD and conformance checking to give you an accurate analysis. In performance analysis, the aim is to find out how well your employees are using the systems prescribed and whether the process was properly designed.

Remember that processes are designed to help you improve operations. Within the event logs, you will find vital data, which will be used to determine how you are performing. For example, you can see how long it takes an employee to perform a given task. A few other employees perform the same task faster. From this, you can find out why the others do theirs faster and why the other employee is struggling to meet the same timelines. By addressing this disparity, you will make changes that affect output. You can also see from event logs whether the changes you made are implemented and if the lagging employee has made any progress.

Process Mining Tools

Based on the different use cases, you can implement different principles and techniques of Lean Six Sigma and in the process improve your business operations. The following are some tools that can be useful when you need to implement specific processes in process mining.

Kofax Insight

Kofax is used by a lot of insurance firms, banks, and logistics enterprises. It combines financial and robotic process automation. It is a useful tool that will give you a clear picture of the processes you are operating, and all this in one platform.

Kofax pulls data from different sources within the entity and presents them in a manner that is easy to analyze. One of the best things about Kofax is that anyone can see their performance record in the same way management will, so it is a transparent approach.

QPR Process Analyzer

This process analyzer is one of the comprehensive and complex products in the market and is backed by brands like Nokia. It is designed to pull data from different solutions like Oracle, Epicor, and Microsoft Dynamics. Some of the features built into it include lead time analysis, automatic process visualization, conformance analysis, KPI analysis, and root cause analysis.

Process Sheet

Process Sheet takes a different approach to process mining. Instead of following the ABPD way, you create templates for processes and use the templates as checklists. This is a good way to create checklists for each task before it is performed. At the end of the task, you can check to see whether it was completed as planned or if the employee veered off track.

One of the perks of using Process Sheet is that it steers your team toward adhering to the process requirements. Once you create a template of what is to be done, the employees must follow the laid-down guidelines, complete with file uploads, form fields, and conditional logic where applicable. Therefore, other than conformance checking, Process Sheet takes things a notch higher by enforcing conformance.

Celonis

Celonis is one of the tools that is taking advantage of the Internet of Things, implementing machine learning in Lean Six Sigma. While other tools might try to establish a procedure and tell you whether the team is conforming, Celonis attempts calculating the odds of decisions on your behalf.

All data that is used in Celonis is presented in a way that you can easily interpret and consider the recommendations offered for improvement. Some of the top brands that are using Celonis include Uber and Exxon-Mobile.

How to Respond to Process Mining Results

While data is important to your processes, data is not the only thing you can use to improve them. You must combine data and astute management practices to improve your operations. Data provides a deeper understanding of what your people are doing and how you can assist them to do more, and efficiently.

First, ensure your employees are part of the iteration and creation meetings that define the projects they will belong to. This is an easy way of giving them leeway into ownership of the projects. It gives them a sense of belonging and can help you smoothen interactions before, during and after the project. Remember that Lean Six Sigma focuses on collaboration.

By collaborating effectively, you will succeed at creating or adding value to processes, which is ideal for your bottom line. The following are tips on how you can use the results from process mining to improve performances:

- **Actionable Processes**

The process mining tools should not be restricted to management only. Make sure they are available to the teams on the ground. The results should be actionable in that they can follow recommendations without supervision.

Employees find it easier to assimilate into processes where

steps are clearly laid out, and they can follow through automatically. To support this cause, consider providing checklists where once they enter records, the information is shared with you, which is a good way to implement documentation.

- *Encourage team buy-ins*

Study your team to see whether they are conforming to the processes you recommend. If you realize that conformance is low, the possibility might be that the team does not share the same sentiments regarding the process as you do, especially regarding importance. This is something you should look at because it could be a sign of problems at different levels.

One of the possible risks could be that your team, people who have done the job for years and have in-depth knowledge of it, do not see your process working. Or if they do, they are not bothered to give it a try. You might also be dealing with a situation where people consider shortcuts to the point where it becomes the norm.

When faced with these challenges, you should not overlook the fact that you are presented with an opportunity to improve services and make sure that everyone sees the benefit in the changes you are proposing. You might not always be right, so where possible, you can consider the recommendations of the employees, and try to get on the same page, instead of fueling a mutiny inadvertently.

- *Efficiency through automation*

Automation is one of the easiest ways of improving efficiency at work. Some mundane tasks can be automated, so you do not have to spend time on them. With a clearly defined process, you have a good chance of realizing where bottlenecks are present and the effect that they have on your system.

- *Process optimization*

The best thing about using data is that you have room to experiment before implementing wholesale changes. Resist the urge to push wholesale changes to your employees. Instead, adjust the systems gradually while running tests on procedural changes. Once satisfied with the output, you can push the changes to the entire department. Continuous improvement is one of the tenets of Lean Six Sigma, which has been standard practice for companies like Toyota over the years.

STRATEGIC CONCEPT OF LEAN SIX SIGMA

Modern Lean Six Sigma is built around some important concepts, which have carried the management philosophy through the years.

The following are the key concepts of Lean Six Sigma that you will learn about:

- Strategy
- Customers
- Variation
- Scientific Investigation

- People

Strategy

Since the inception of Lean Six Sigma, the concept has always been built around quality assurance. At tactical levels, quality assurance is mandatory with a view to completing projects on time and reducing costs in the process. Therefore, all strategies that can be employed to ensure this kind of continuity should receive full support from the structures and processes within the organization to make sure the business moves forward in its goals.

In Lean Six Sigma, the organization should have separate factions that determine the strategic objectives to be achieved and another that determines the definition of projects. From project definition, you should also have feedback to the strategy team so that both of the teams can appreciate their contribution to the overall objectives of the company and what this would mean at the highest level.

In any organization, senior management often offers their support to the projects. However, in a practical sense, most of the work is passed down to quality assurance managers, and even with their focus on the obvious benefits of delegation, the lack of a centralized decision-making approach becomes a problem

Through Lean Six Sigma, focus on projects is more on the quality teams because they are in direct contact with the staff and elements that are involved in the processes. It is important for management, therefore, to take a personal interest in the Lean Six Sigma training schedules and interact with their teams as they undergo training, answering some of the questions that might arise.

Another approach to strategy is to get reviews on the

process and how people are adapting to changes over time. You can get monthly or even weekly discussions where you review the procedures implemented, discuss some of the difficulties that people are having, and address any issues arising.

If you work for a company that has multiple sites away from the administrative center, make a point of visiting the individual sites to see how Lean Six Sigma training is being integrated into your business culture and how this has affected the operations. You also have to set time aside to discuss the reports and reviews with the top brass in Lean Six Sigma hierarchy, the master black belt team, if you have some. Lean Six Sigma will give you a significant cultural impact if you can aptly combine direct involvement and commitment with senior management support.

Customers

Bad customer service is more common than you can imagine. It is something that companies grapple with all the time. Lean Six Sigma is a quality management approach that can help managers learn how to do away with inefficiencies and processes that result in poor customer service.

Through Lean Six Sigma, you can come up with a disciplined approach that will help you focus on an almost perfect delivery for services to your customers. You must identify the factors that are important to your customers, especially those that affect the quality of service or product delivery to them. How can this be done?

First, you must define what customer service means to you. Once you understand the problem you are dealing with, you can address it in terms of the challenges involved and how to meet the desired outcome. You must also understand what problems you are having with customer services. Can you establish how the problems come about? Why are you having those problems?

Building on your understanding of the current customer problems you are solving, you must also understand what is happening in terms of the variables. What or who are involved in the problem you are dealing with? Where possible, employ an evidence-based approach to answer this question. With evidence, you have a lot of information at your disposal to help you map a correct plan of action.

You can then analyze the details of the problem you are dealing with and how they come about. Look at the circumstances under which you encounter customer problems. Are there specific factions in the company responsible for these problems? Is there a way for you to address the problems amicably?

It is from here that you can then find a way to improve customer services. Given your experience in Lean Six Sigma, you will have recommendations that you can share with the top brass in management on how to handle the issue with customers. Remember that customers are integral to the success of your business; hence, their satisfaction is important. Even when addressing the customer needs, however, remember to stay objective in addressing customer satisfaction so that you do not end up veering from the company traditions in a bid to please the customer.

Whatever you do to improve the customer experiences with your company, it must be subject to frequent evaluation. Did it work for your immediate case? Has it been working since then? How can you ascertain that similar problems will not arise in the future? Is there a way you can replicate the same results to other aspects of the business?

One of the challenges that most companies have is the

assumption that they are fully aware of what customers want. This is far from the concept of customer focus. In customer focus, you must ask customers what they value, what they want, and how they rate what you are currently offering them. At times you do not need to ask directly. Take the case of Apple, for example. The company had for a long time stuck by their guns, insisting that they would never make devices with gigantic screens. However, the Android market had embraced; this and customers were getting used to devices with gigantic screens. While other companies would have wasted a lot of time trying to find out what the customers wanted, Apple studied the trends in mobile computing and decided to adapt to the market needs especially for their flagship products.

In Lean Six Sigma, you must recognize the value that your customers bring to your business and, in return, create value for them. Even while you are doing that, you must also focus on reducing costs where necessary while still offering profitability to your customers over the long term. The emphasis should be on keeping customers happy for the long term, over short-term financial benefits.

Variation

Variation is a concern in most production processes. You will learn in Lean Six Sigma that product variation can introduce a lot of problems, like defects. Other than defects, variation also reduces the level of customer satisfaction and comes with added costs. In the long run, variation further affects your expected revenue in the future.

Varying components can create an issue in assembly lines, especially when the variation is to a notable degree. You end up wasting time and other resources making adjustments and fittings to correct the variations. This is time that would have been saved if perhaps the assembly process was uniform.

Inconsistencies in product performances are the same factors that will introduce dissatisfaction to customers, and over time, their purchases will reduce. If you have products that are constantly performing poorly in the market as a result of variation, these products will not just have a bad reputation, but the reputation of the company, in general, will take a hit.

The more intense your desire is to differentiate products, the riskier it will get for you in terms of waste created and the probability of errors. It is easier to identify errors and fix them in a simple unitary system than addressing errors in a system that is strung up in product variations. Whether your company is in the manufacturing sector or not, variation will cost you.

Scientific Investigation

Quality management methods are aimed at improving processes. From Lean Six Sigma to Agile, they all must establish a hypothesis, test the hypothesis, and continuously adjust the systems to accommodate their findings. This is the same approach that is used in modern strategy management and product development approaches. Science, therefore, must be accorded the necessary room in management.

Lean Six Sigma thrives by moving companies toward the use of factual information built on data. It takes away the element of relying on the leader's whim to make decisions and, instead, uses trends, analysis, and tangible data.

Before a project begins, you must have a clear goal and an understanding of how to determine whether you have achieved the goals at the end of the project. From there you come up with plans on how to achieve what you are setting out to do. The plans must then be acted upon, and the guidelines around the plans observed.

When the project terminates, you must analyze the results and compare the current position with the earlier position to determine whether the original plan should be modified or not. This is all a scientific process, and it is the heart of Lean Six Sigma. Like any scientific process, the following cycle must be keenly followed if you are to succeed with implementing Lean Six Sigma in your company:

- Objective observation. You must take a keen interest in the subject of your study or project to understand what it is about. If it is a problem that needs to be addressed, you must study it keenly to understand how it manifests, and what the effect is.
- Formulate questions that define your observation above. This question defines the problem you are facing. It might be a difficult process, but it should be done well as it provides the foundation for your hypothesis.
- Build a hypothesis. The hypothesis should not be limited to describing what you have observed, but should also address your expectations for the future if you succeed in building an improved solution for the problem. A good hypothesis must always have a predictive value.
- Conduct tests to determine the feasibility of your predictions.
- Evaluate to see whether the hypothesis is succinct. Did you improve the processes you set out to? Did you address the problems that got you to this point in time?

Depending on the result of the evaluation, you can modify

the hypothesis and address another problem altogether. This scientific cycle persists until you have addressed the issues the company has been struggling with.

People

The biggest challenge to any organization in transformation is always people management. Even in security, the consensus is that human interaction is the weakest link in the strongest security system. The individual performances of the employees often determine whether the business will succeed or fail. People form an integral part of any quality system. They can help it succeed or bring it down.

In Lean Six Sigma, you must incorporate frequent training and assessments to make sure your employees meet the requirements for your success in the market you operate in. Skill evaluation is mandatory because it is from there that you will identify the talent gaps and work on how to fill them.

Where people are involved, you must also introduce the element of learning. People improvement is an approach in Lean Six Sigma that confers significant benefits to the company in the future. Each company has its traditions, upon which culture is conceived. It is the culture in the company that keeps employees committed to working with the organization for long.

Without getting the majority of the employees to accept and appreciate Lean Six Sigma, it will be difficult for them to go beyond the bare minimum and realize their true potential. The result here will be hurting the expectations of the company as a whole.

When implementing changes in the process of people improvement, you must anticipate resistance. Do not mistake resistance to change as a big problem but a natural reaction. You can address the reasons for resistance and listen to them. More often, they have valid reasons for resistance. The mistake that most companies make is looking at those who offer resistance as a problem for the organization and castigating or punishing them. A situation like this creates an *us-versus-them* conflict.

Inclusivity is important, and an effective strategy that can help you succeed in your business whether in the public or private sector. It is also an important aspect of Lean Six Sigma that will help you manage your personnel better and enjoy good returns in the future.

A lot of people look at Lean Six Sigma from the perspective of financial gains and benefits. However, they ignore the role that people play in Lean Six Sigma. Focusing on financial gains poses a challenge to efficient implementation of Lean Six Sigma in any organization. This is why most companies end up using Lean Six Sigma for cost reduction. The problem with this approach is that it is single-minded and you will barely achieve what you want in the end.

4

DMAIC

Process improvement protocols like Lean Six Sigma promise a lot of benefits. However, how do they work? What is the driving force behind them? For Lean Six Sigma, DMAIC is one of two core techniques that you will implement. It is an approach that will get you from a beginner to a master in Lean Six Sigma.

DMAIC is an acronym that stands for:

- Define
- Measure
- Analyze
- Improve
- Control

It is an improvement cycle that is built around data, and when implemented in your business, it will help you identify the inefficiencies that cause defects in output and learn how to get rid of them. The idea behind DMAIC is to optimize, improve, and make your existing business processes stable (Shankar, 2009).

Assessing Suitability for DMAIC

Before you decide whether DMAIC is the perfect tool to address your needs, you must run an evaluation test. Some companies refer to this as Recognize and use it as the first step before DMAIC.

Recognize is not addressed as a step in the DMAIC process, but it serves a purpose. The reason for this is because DMAIC will not always be applicable in all situations. It is only suitable to situations where you need to improve processes.

For a situation to be suitable for DMAIC, it must meet the following three conditions:

- Imminent problem

The problem must be obvious and supported by a set of processes or an existing process.

- Improvement

You should see room for improvement, especially when you reduce or eliminate defects or lead times and other variables, while at the same time improving the profitability of the business through cost savings or improved productivity.

- Quantifiability

The issue you are addressing must be quantifiable. Find out whether you can measure the data or if the results are easy to understand in a quantifiable manner.

If the project meets these requirements, you can map it onto DMAIC.

Define

At this stage, your role is to understand and plan the exercise accordingly. You must understand the requirements of your customers. To do this, understand the type of customers you have, internal or external.

Internal customers are management level within the company or any other departments that depend on the output you are producing so they can make decisions for their roles. External customers are those end users who pay for your products and services. Other than business clients, they can also be shareholders in the company.

The following are steps you will follow in the Define stage:

- Define customers and customer requirements.
- Establish a problem statement based on your goals and the benefits expected.
- Identify the team, process owner, and project champion.
- Define project resources.
- Identify the important organizational support for the project.
- Establish a project plan and applicable milestones.
- Create a process map.

To meet the requirements above, the following are the resources you will need:

- SIPOC diagram
- Project charter
- Process flowchart
- A work breakdown structure
- Customers and requirement definitions

Measure

How is the process you want to improve performing currently? What you are trying to find out here is the magnitude of the problem at hand. Collect data on the process, but at the same time, make sure you are sure about what the customers want.

The measure step is all about data. You must identify defects, opportunities, and the metrics aligned with them. You must be keen on the data collection approach because the ease of translation will affect the outcome. You must have a standard collection process to help you get better results.

The procedure for measuring to determine the current performance level and quantify the problems identified include:

- Identify opportunities, defects, metrics, and units.
- Obtain a process map detailing the affected areas.
- Establish a data collection plan.
- Ensure the measurement system is credible.
- Data collection.
- Establish a $Y=f(x)$ relationship
- Establish a sigma baseline and process capability

The tools you will need for this stage include the following:

- A process flowchart
- Process sigma calculation
- Measurement system
- Benchmarking
- Data collection example or plan

Analyze

The demands of this stage are to help you understand what

might be the reason behind the problems you are facing. Most people do not give this stage the attention it deserves and end up jumping to presenting solutions before they know what the real cause of their problems are.

When you end up with solutions to problems you do not understand, you will repeat the same process each time you have a problem. It is best to understand the cause of the problem so that you implement solutions that tackle these issues from their core.

Your task is to find out the real causes of the problems here, not the solutions. You need a hypothesis addressing why the problems exist and then work toward disapproving or proving the hypothesis.

The analyze process involves the following:

- Determining the important x's in the Y=f(x) relationship.
- Establishing the root causes.
- Establishing the reasons behind the variation.
- Identifying and separating value and nonvalue addition steps and processes.
- Defining performance objectives.

You will need the following tools:

- A Pareto chart
- A histogram
- Scatter plot
- Run chart (time series chart)
- Fishbone diagram (cause/effect) chart
- The Five-Why approach
- Regression analysis
- Statistical analysis

- Process review map and analysis
- Continuous and discrete hypothesis testing
- Data analysis for non-normal data

Improve

Having identified the problems the company is dealing with, the team must come up with feasible solutions. This stage calls for a lot of brainstorming sessions. Proposals will be made for possible changes, and at the end of this process, make recommendations on possible causes for improvement. The recommendations should be innovative if they are to meet your desired goals and eventually improve the experience customers have with your products or services.

The improve phase includes the following:

- Carrying out random experiments
- Establishing prospective solutions to problems
- Defining operational tolerances in the prospective system
- Validating potential improvements by conducting pilot studies
- Assessing room for failure in the recommended solution
- Evaluation and re-evaluation

You will need the following tools:

- Pugh matrix
- Experiment design
- Mistake proofing
- Brainstorming
- Simulation programs
- FMEA (failure modes and effects analysis)
- House of quality (QFD)

Control

Having made recommendations, you must ensure that they are sustainable. The control stage comes after the problems have been resolved and all the improvements desired are in place. The gains achieved by implementing this process must be maintained.

To achieve this, you must come up with a plan for monitoring—a plan that ensures the success of all the processes—and should there be a slump in performance, you must ensure you have a rescue or response plan ready.

To control performance in the future, this phase will involve the following:

- Definition, validation, monitoring, and control systems.
- Establish standards and procedure.
- Determine the capability of the process recommended.
- Implementing statistical process control approaches.
- Create a transfer plan that will be used when handing over the process to the owner or senior management.
- Verify the cost savings, benefits, and profit growth before presenting to management.
- Communication with the business.
- Present the final documentation and close the project.

During this phase you will need the following tools:

- A control plan
- Cost savings calculation
- Attribute and variable control charts

- Process sigma calculations.

Features of DMAIC Programs

DMAIC programs are designed to help your company identify and enjoy utmost utility from your improvement efforts. The following are some of the key features you should be looking for:

Improvement Opportunity

Form the moment any of the employees identify an opportunity for improvement in the production cycle, the improvement cycle starts. An effective DMAIC solution should be one that eases the process such that any employee can recommend an opportunity whenever they identify it.

An efficient program should also allow the employees to upload images, documents, and any content that might be useful in evaluating the opportunities. An easy submission process helps you overcome engagement barriers that are often associated with most bottlenecks. Remember that your staff are busy going about their roles, so if the process is too difficult for them to utilize, they probably will not participate.

Workflow Activity

From the moment you initiate the DMAIC process, you must maintain the same momentum and ensure you do not miss critical tasks. Good DMAIC software can send you notifications and alerts to make sure everyone is aware when their input is required. The program should also have a dashboard where you can view progress and stalled processes. Today you have programs that have collaboration features to keep the whole team working in sync.

Reporting Impacts

Each improvement recommendation should have an impact on the overall process. You, therefore, must be in a position to calculate the impact that the improvements have. You need to know whether the changes expected are qualitative, financial or if the system is indifferent to change.

This reporting is important especially at the management level because it allows them to determine whether your project needs more support or not. You must also communicate the impact to employees so that they are motivated to keep working toward the ultimate goals.

Capturing Knowledge

The DMAIC cycle does not end. It is repeated with improvements on previous levels over time. You must ensure that the knowledge you learn from each improvement cycle is stored and can be used in the future to assess historical gains. This knowledge can also help you identify trends and how they can be used for future decision making.

5

DMADV

DMADV is one of the Lean Six Sigma frameworks that is built around developing new processes, products, or services. This is the fundamental difference that sets it apart from DMAIC, that focuses on improving existing processes. The DMADV process is useful when you are implementing new initiatives or strategies because it is built to use data in identifying success projects and in-depth analysis.

Where is it applicable? DMADV is useful when you do not yet have a product or a process. It is perfect for companies that are just starting out. Apart from that, DMADV is also useful in existing processes or where the product or service is already in the market, but there is a need to improve customer satisfaction or meet set Lean Six Sigma requirements.

DMADV is built into five phases as follows:

- Define
- Measurement
- Analysis

- Design
- Verify

Define

DMADV begins by determining the reason behind the process, service, or product. From here, you determine the need for and set measurable and realistic goals that are aligned with the objectives of the company or the stakeholders.

In this phase, you must also come up with set guidelines and schedules within which you will identify, review, and assess any risks in the near future. The define phase is integral because it is here that you establish the definition of the project and all the strategies and goals that are aligned with the company expectations and those of the customers.

Measurement

What are the factors that will influence the quality of the product, service, or process? These factors are measured in this phase. You should define the market segment and requirements thereof, the parameters mandatory for effective design, scorecards that will be used to evaluate the components of the design, and reassess the production process and any risks imminent.

Having figured out the values associated with these factors, you can come up with a practical approach to initiate the process of production. You must also make sure you are clear on the metrics that will be given priority, especially those that are dear to the stakeholders. This is important because these are the metrics that eventually translate to the expectations and requirements that customers have of the output, and they become your project goals.

Analysis

The analysis stage builds on the data that was collected on the measurement stage. Some of the activities involved in this stage include determining the ideal combination of requirements that will deliver the best outcome or components, how to establish the right value considering the constraints present, and how to develop the best design.

During the analysis stage, you must also determine the cost of the design spread across the life of the project. You should present a number of options and, upon consideration, settle on the best option to meet the desired goals.

Design

The design stage requires professional insight in designing the right approach for the alternatives you choose. Priority is given to the elements that make up the selected alternatives, and it is from this that you can develop the best design that will take the company forward.

Building on this, you will work on a detailed model. It is the detailed model that eventually determines where errors might be present and how to make changes to accommodate room for improvement.

Verify

By this point, a design must have been presented, which is admissible to stakeholders. You research to find out whether the prototype presented is applicable given real-time metrics in the real world. You will also initiate a number of production runs to pilot the project so that you can determine how it performs and to make sure you get the best quality out of it.

It is during this stage that you can confirm whether your expectations have been met or not. All the findings you get from the pilot stage must be documented, because they will be used to determine a feasible deployment plan. Another important subject that you must discuss is how the product plan will be transitioned from the pilot stage and turned into a routine part of the production process and to ensure that the changes that come with it are sustainable.

While there might be slight disparities in the emphasis especially concerning different components of the DMADV approach, the overall objective still remains the same, making sure that the product, process, or service delivers quality results that can be replicated and sustained over the long-term in the face of normal business operations and constraints.

DMAIC and DMADV: Similarities

Both of the methodologies use statistical approach and tools to identify and propose possible solutions to problems the company might be having in quality control and assurance.

To incorporate either of these strategies in the workplace, you must have attained at least a green belt certification.

Your ultimate goal must be reducing waste in the production process.

All the solutions presented in either case revolve around data usage, and they are backed up by facts.

DMAIC and DMADV: Differences

The apparent difference between these tools is how they adapt the penultimate steps. While the DMADV process redesigns a process to make sure it is suitable to the customer requirements in the design and verify face, the DMAIC process does it differently.

In DMAIC, the final two processes are geared around process control and adjustment. It is concerned with defining the business processes and how adaptable they are. The DMADV process, however, focuses on the relationship between customer requirements and the services or products at an output.

DMADV uses a business model to measure the efficiency of the proposals. In this case, you have to run simulations to determine the efficiency of the business model proposed. DMAIC, on the other hand, uses a control system to determine how the business entity will perform in the future.

DMAIC and DMADV: Application

Before you choose between DMAIC and DMADV, you must understand the nature of the project you are handling. Business needs take precedence, and you must have a system in place that will help you figure out the methodology that is suitable for your value addition goals.

For a company that is introducing a new project, process, or service in the market, DMADV is a suitable alternative. In case the project is already existing, DMAIC should be suitable. Some Lean Six Sigma experts do recommend the use of DMADV in a situation where the process improvement has not met any expectations.

In the case of DMAIC, it is recommended as an improvement protocol. In relation to customers, DMAIC is ideal in a case where you are dealing with products that do not meet customer needs anymore. It is useful to help you gain customer trust once again and belief in your business process.

Since most people can barely tell these two philosophies apart or where they should be applicable, it is wise to bring on board an experienced Lean Six Sigma expert. A master black belt would be suitable in that they can help you make a decision on the best method to apply.

Lean Six Sigma is among the best and powerful quality management tools and methodologies that you can implement in the company to meet your customer expectations and needs. However, it is not always easy to determine the better option between DMADV and DMAIC. Given this challenge, you must have an in-depth understanding of your current business processes, because with this knowledge, you can decide how you want to proceed, without diverting from the company goals.

6

COMPARING LEAN AND LEAN SIX SIGMA

I f you are just starting out in Lean Six Sigma, you might get confused when you come across these terminologies. It is advisable that you learn the differences between them so that you can tell them apart and, more importantly, so that you can learn which one is ideal for your situation. Lean, Lean Six Sigma, and Six Sigma might share a few similarities, but they are different procedures altogether (George, 2002).

Six Sigma

Six Sigma is about process improvement. We can trace the history of Six Sigma back to the Motorola manufacturing plant. From the gains by Motorola, this methodology has been adopted in many companies and industries, especially in the manufacturing sector. All companies strive to enjoy continuous improvement in their business processes.

The primary focus of this methodology is to reduce errors and variances within the individual company's production process. The persuasion behind this is that if you cannot depend on the business process for consistency, the final outcome will be an excess of damaged or defective products at an output and many errors that are a waste of company resources. To improve the quality of delivery of products and services, therefore, you must reduce the current variance levels.

The Six Sigma standard that you must aspire to achieve is reducing defects to 3.4 for every 1 million opportunities accorded. Six Sigma proposes two methods to achieve these goals. These methods are all factual, data-driven, and are necessary to identify and eliminate waste in the business processes. These methods are as follows:

- DMAIC (Define, Measure, Analyze, Improve, and Control)
- DMADV (Define, Measure, Analyze, Design, and Verify)

DMAIC is vital in identifying and correcting issues present in your current business processes. DMADV, on the other hand, is useful in creating new processes.

Lean

Lean borrows a lot from Six Sigma. It is a methodology that is hinged on improving business processes. However, the only difference is that instead of identifying and eliminating waste from the process, Lean attacks waste.

Lean manufacturing focuses on eliminating waste to maximize your output value to customers without compromising on the productivity quality. Lean quality management can be traced back to the Toyota Production System in Japan between 1948 and 1975. The Toyota Production System was built to prevent the following three things:

- *Muda*

Anything within the manufacturing process that introduces constraints or creates waste, affecting value addition to your products.

- *Mura*

Anything that brings about inefficiency or inconsistency with workflows.

- *Muri*

Loads or tasks that add unnecessary stress to your resources, especially machinery and your employees.

Other than these, the Lean manufacturing system was built around the following principles, that have guided organizations over the years:

- *Value*

The company must do all it can within its power to deliver nothing but valuable products to the customers.

- *Value stream*

There should be clear, laid-out procedures that will be used in delivering quality, value-adding products, and services to customers.

- *Flow*

Make sure that all the steps that are necessary for value addition are running smoothly without any bottlenecks, delays, or interruptions whatsoever.

- *Pull*

Production should be on a *just-in-time* structure to make sure that you are not stockpiling materials. As a result, customers should not have to wait for months to get their products when they can get them in weeks.

- *Perfection*

The Lean methodology should be inculcated into the company in process improvement so that it becomes a tradition.

In this methodology, anything that is not adding value to the customers, whether in the service or production industry has to go. With this in mind, Lean addresses waste elimination in the following main areas commonly referred to as the deadly lean wastes.

- *Process defects*

Process defects refer to mistakes in the production or service processes in the organization, which will consume more resources to remedy. In the case of manufacturing, this might refer to a defective part, for which you must commit more funds to remake.

Process defects are common in organizations because of undocumented changes in design, improper design, lack of control in the inventory levels, missing or weak processes, inappropriate documentation, lack of quality control, or neglected repair.

While it might not be possible to achieve 100 percent elimination of waste, you can reduce defects in the organization by carefully standardizing the production procedure. You can also institute astute quality control measures in business processes and make sure that all the employees know their roles and the importance of their roles to meeting customer requirements.

- *Overproduction*

One of the challenges most organizations go through is overproduction. There is no need to keep churning out products when either the end user does not find value in it or they are not ready for the products. The problem that this creates is tying up working capital unnecessarily. The manufacturing industry is notorious for overproduction, though as long as any organization suffers bottlenecks that are yet to be identified, overproduction is always imminent.

Some of the reasons you might be overproducing include depending on a forecast, changes in your engineering processes, production on a just-in-case manner, lack of proper understanding of the end user needs, and automation without proper planning.

To combat overproduction, you must create a workflow that is reasonable enough and aligned with the needs of the customers, while still meeting the company objectives. You must also make sure that the company has set procedures in place that will streamline all processes within the organization. If possible, you can also recommend processes that can prevent backlogs against bottlenecks that have already been identified.

- *Idle time (waiting) between operations*

This is a problem that manifests when you need to stop work for any given reason. A possible explanation for this predicament could be inaccessibility to all the materials you need, you are waiting for approval, or someone is overwhelmed by their job description.

Some of the reasons responsible for an idle time between operations include lack of proper communication, poor process quality, absenteeism at work, insufficient staffing, and an unbalanced workload.

The problem with idle time is that an efficient employee or process will always have to wait for specific bottlenecks to clear. The easiest way of addressing this issue is to ensure you have the right staffing procedures, relative to the workload experienced in processes that have identified bottlenecks.

- *Nonutilized talent*

Nonutilized talent is a big concern in most establishments today. When you hire people but you are unable to utilize their skills, talent, or knowledge, you end up with waste. Most companies struggle with this because they are unable to understand the people they bring onboard.

The problems with unutilized talent can manifest in the company in the form of poor management, lack of proper training, friction in team assignments, lack of efficient communication, wastage in administrative duties, and assigning the wrong roles to employees.

The problem with nonutilized talent is that this is often the most significant cause of disengagement by employees. Any organization that is struggling with such challenges can consider taking an interest in the employees, getting them trained and reducing micromanagement.

- *Wasteful transportation ethos*

The logistics behind transportation can cause a lot of problems for the organization. In most offices, this is barely a concern. However, in a manufacturing plant, transportation logistics are a severe issue. In an office, the cost and challenges of transportation can be mitigated through emails. The problem with transport is that too much motion increases overheads. There is also the risk of product damage, lack of communication, and product deterioration.

Transportation wastes are often brought about by inappropriately designed systems, a misaligned process flow, the presence of unnecessary steps in the transport process, or a crude layout in the manufacturing plant.

Managing transportation waste in most cases requires common sense. Look at ways of simplifying the transport crisis. Where there are physical changes to be made, make them. Try to make sure that you spend less time dealing with processes, and instead, shorten the distance between steps.

- *Excess inventory*

Inventory excesses come about when you produce beyond your customer demands. This raises the problem of masking the real production. Some of the causes of inventory excesses can be attributed to misunderstanding the needs of your customers, having lengthy setup times, using unreliable suppliers, inappropriate monitoring systems, and creating unnecessary buffers, hence overproduction.

- *Unnecessary machine or employee actions*

Any movement that is surplus to requirement sets you back, whether it machinery or employee motion. Common causes of unnecessary motion include sharing machines and tools, a congested workstation, operating without standards, isolated operational facilities, or process control and design. Managing this problem can be as simple as redesigning the workplace so the layout is more efficient, especially if you can reduce the distance between stations. In the redesign, try to make sure your team can get easy access to all the tools and equipment that they need to use often.

- *Nonvalue adding processing*

Excess processing is a challenge in most companies because they end up creating different versions of the same job. As a result, the company has to process a lot more than they need, especially in light of processes that are not designed properly.

Common cases of excess processing include human error, lack of understanding in lieu of the customer requirements, lack of communication, duplication of data, roles and processes, demanding unnecessary signatories, and excess reporting procedures.

All these processes drive up your operation costs. They are resource-intensive, and you will waste a lot of time on them. To deal with this, you must take the time to study the company and analyze all the processes so that you know where waste is imminent and fix it. Processes that can be standardized should be. Do away with unnecessary documentation by empowering team leaders to sign off on meetings and processes instead of waiting for managers.

A careful glance at the wastes above can be summarized as DOWNTIME (defects, overproduction, waiting, nonutilized talent, transportation, inventory excesses, motion waste, and excess processing).

Lean was initially built to improve manufacturing processes. However, the methodology has proven useful in so many industries. Instead of concentrating on a single point in any process, Lean optimizes the process through all the unique assets, technologies, and departments until it gets to the end user.

The fact that Lean focuses on all the processes instead of an isolated moment in the process flow helps to do more than waste elimination. Instead, Lean helps to inculcate a culture of continuous improvement across the organization. Lean has also been touted as a very good way to motivate and empower employees, given that they can see and associate with change as it happens. In the long run, the dedication to waste elimination methods as advocated in Lean results in an efficient operation, better product and service delivery, and improved yield.

Difference between Lean and Six Sigma

- *Statistical versus philosophical approach*

The approach used in these processes accounts for one of their dissimilarities. Six Sigma uses a mathematical, quantitative approach that is leveraged on data to champion for change in the company. In Six Sigma, you must concentrate on a single element in the business process and work toward improving it before you can move to another section and improve it and so forth. It is a step by step procedure.

Lean, on the other hand, addresses your concerns through a company-wide improvement. Lean focuses on wholesale changes carried out over time. The core objective is to make sure that the operational flow is smooth in all sectors, in the process improving your productivity and reducing or eliminating waste. Lean, however, does not give you a specific time within which you must complete the improvement process.

- *Problem solution versus continuity*

Six Sigma is built around solving problems that are setting you back. You identify problems and assign a specific timeline to find solutions, and fix them. All the resources the company makes available for Six Sigma are deployed to the problem until you solve it.

Lean consists of a wide array of methods that are applied to improvements on a daily basis. It is a process of continuity, also referred to as the *Kaizen approach*. The methods that are used in Lean are applicable to any manufacturing process and equipment and can be used to eliminate all kinds of waste in the company.

You will notice that Six Sigma proposes a data-driven approach that recommends unique improvements for individual projects. On the other hand, Lean focuses on general improvements that are carried out on a daily basis.

- *Top-down versus bottom-up approach*

Six Sigma is built around a hierarchy. The structure revolves around different belt levels, with each level representing unique capabilities and knowledge of tackling issues at that level.

Lean, on the other hand, does not propose a hierarchical structure as in Six Sigma to recommend and implement changes to processes. Everyone in their capacity, their roles in the operation notwithstanding, should be an active agent in improving the management process.

The distinction between Lean and Six Sigma is that in the Lean approach, anyone can notice a problem and instruct the rest of the team on what they should be focusing on and even recommend a method that they consider suitable for this.

This is a bottom-up approach where anyone can experiment and test for you to see the strategies that will work for your company.

With Six Sigma, however, you must get a qualified expert to come in, structure a suitable team for the problem at hand, and manage that team until they meet their goals. This is a top-down approach where problem-solving is left to professionals. They decide when the problem has been solved and the project completed.

Lean Six Sigma

Lean Six Sigma is a hybrid methodology, and like all hybrid methodologies, it combines the best of both worlds. Lean Six Sigma was proposed when companies realized that these two methodologies could be combined to eliminate issues that were arising in areas where the two were deployed independent of each other.

If you are using Lean Six Sigma in your company, what you are trying to do is to get rid of waste through the Lean processes and, at the same time, introduce DMAIC and DMADV to improve your business processes. An astute combination of these processes is useful in that it will help your company become efficient in all the operational processes and, at the same time, make sure that your output in products or service delivery is unmatched.

So how do you choose which of these three to implement in your organization? Instead of choosing Lean, or Six Sigma, it makes sense to choose Lean Six Sigma. The choice, however, will depend on what your company needs. You must conduct a needs assessment to know what ails your business processes before you can recommend an appropriate approach.

Irrespective of the differences in the approaches used, all

these methodologies are geared toward improving your competitive advantage. With the right training, you have the necessary skills champion change in the organization and implement the recommendations that will make you more competitive in the market.

Identifying the Right Method for Your Company

People often debate about which of these quality management approaches is the right one to implement in your organization. However, Lean and Six Sigma are all geared toward identifying and getting rid of inefficiency within your business processes. Perhaps the main distinction between these methodologies is the approach used, but the desired outcome is the same.

Before you begin, you must realize that the method and approach you use will affect your business at all levels. Depending on how you implement it, the organizational processes will face certain implementation challenges. This is why master black belt experts must be agents of change. They have the technical capacity to manage the transition and ensure less friction in the process.

Conduct an honest needs assessment to determine what your company needs, and ensure they go hand in hand with your strategic objectives and goals of your business. Remember that what works for one company will not necessarily work for another. For utmost flexibility, experts recommend Lean Six Sigma over implementing Lean or Six Sigma in isolation.

While the methodologies presented might have been around for years, there have been many changes to them over the years. With this in mind, be ready to evolve and stay flexible in implementation. Your business description and needs might change from time to time. With an able master black belt at the helm, these changes should be accommodated in your approach without any concern.

Let's take the example of a company that starts out as a software company. Over the years you will implement different Lean Six Sigma recommendations to streamline your business operation. However, a few years down the line, your company grows, and you venture into hardware products. You will need to make changes to things like inventory management, internal communication, customer services, and so forth. Such changes would indicate a significant shift in the company goals, but with a professional in charge, all should be well.

7

UNDERSTANDING CUSTOMER NEEDS

C ustomer satisfaction is not one of the easiest things to achieve, let alone maintain. The modern customer has dynamic needs that change with respect to different stimuli in their immediate and external environment.

It is too difficult to understand customers to the point where even some of the companies with the most advanced, customer-centric processes still struggle to understand what their customers want. More often, they realize they have a problem when it is too late, and customers have moved on to other brands.

Since it might not always be possible to understand customers, you need to focus your attention on them to ensure that your strategies are aligned as close as possible with their needs. You need a strong link between what customers demand and your value proposition. This is what you get in Lean Six Sigma.

The concept of Lean Six Sigma starts with and ends with the customer. Everything you do at company level ends up in meeting customer needs. Lean Six Sigma might have been borne out of the manufacturing industry, but the impact, especially in the modern world, is felt in marketing and sales. Companies that have succeeded in implementing Lean Six Sigma over the years have done so by embedding it in the following areas of their businesses:

Client Relationship Management

Client relationship management (CRM) refers to technologies, strategies, and practices that your company uses to analyze and manage customer needs and interactions through the course of the interaction life cycle with your company. The idea behind CRM is to make sure the company enjoys a high rate of customer retention, improve customer services, and support the sales growth needs.

CRM systems used in different companies collect a lot of data at the point of interaction between the company and the customer. This is then used to analyze your preferences and address any concerns that the customer has. They collect and document customer information, giving the business a better shot at access and management.

Today, CRM has advanced, and more tools have been built into them, which enable the company to use the customer's collected data to provide better services to the customer. CRM tools have improved, and now they contain features like sales force automation, marketing automation, contact center automation, and they also allow the company to offer services based on the customer location.

Sales Effectiveness

Sales effectiveness loosely refers to your company's ability to drive sales. A higher sales effectiveness means you are succeeding in the market by addressing the right target audience. This also means that you are working closer to meeting your operational and financial goals.

Sales effectiveness refers to a collection of all the things your business must do to help you succeed. To succeed, you must ensure that there is a concerted effort from all the teams that are involved in the sales and marketing tasks, including operations. There are three ways of driving sales effectiveness:

- *Define*

You must start by defining your key performance indicators (KPIs). Identify those that are extremely important to the company, and outline the skills necessary to meet these indicators. Once you have the KPIs prioritized, you can discuss and design an approach that is suitable for your team, considering the market that you are trading in.

From there, you must ensure the company is aligned with the rudimentary concepts of the strategy you propose, making sure that your sales pattern and behavior is linked to the outcomes that you are working toward.

- *Develop*

In this stage, the sales managers and sales personnel interact to learn from each other. This should be an engaging session where try to apply the sales concepts learned. You should provide useful media content to show the team what they should work with. These meetings should also be about confidence building.

- *Sustain*

In sustenance, you work toward establishing a system where you maintain the progress you have made in terms of training your employees to meet customer needs. Feedback is important for this, consultation and mentorship. This is where your Lean Six Sigma skills will come in handy.

To boost sales effectiveness, you need to eliminate all aspects of ambiguity in the process. The following are areas you must pay attention to, in order to succeed in sales effectiveness:

The sales process. Make sure you have a sales process. Analyze and study it to ensure it is effective, consistently used by your team, and is understood by everyone in the team.

Managing opportunities. Sales opportunities arise, but most companies barely know how to deal with them. Your role as a Lean Six Sigma expert is to make sure you have accurate sales forecasts and use this to manage the opportunities that come your way. You need a sales pipeline with clearly outlined stages.

Sales efficiency. Consider the current transaction size of your company to determine whether you have a reasonable sales cycle. In line with waste elimination according to Lean Six Sigma principles, ask yourself whether you have sales opportunities that barely amount to success but are constantly running in the company.

Sales performance. Take time and evaluate the sales team you have. How long does it usually take before your sales representatives are productive and are turning in results? How many of the representatives are meeting their quotas?

Sales skills. Discuss skills and training with your sales team. This should be an honest discussion, given that what they learn will be useful in managing your customers. Are your sales representatives capable of closing deals, planning calls, and identifying opportunities or priorities?

Developing New Markets

New market development is one of the methods that companies use to widen their scope and strategy. We have too much-unexploited potential in all industries, and companies often struggle to tap into this. Developing new markets is a growth strategy that will help you venture into new territories before the current market is unable to sustain your business. You need to develop new markets to help you survive the difficult times.

When planning how to get into a new market in line with the growth plan for your company, the ideas mentioned below can be useful:

- *Identify the target market*

Your target market refers to the group of customers that you feel are highly likely to be fascinated by the products or services you are offering. To identify your target market, look to your customer profile for guidance.

- *Customer profiling*

A Lean Six Sigma professional should have a deep under-standing of the business, the current customers and the profile of the customers. You should tell at a glance the average age of your customers, the gender, their purchase habits, their marital status, and so forth. These are metrics that help to increase your chance of success when you get into a new market.

- *Market demographics*

What features describe the market you are going after? What is their income level, age, growth rate, the size of the market, the purchase habits, and so forth? Market demographics and customer profiling go hand in hand. A good understanding of your market demographics is crucial in determining whether the target market is sizeable enough to meet the goods and services that you are offering.

- *In-depth market analysis*

An in-depth analysis of the customer market will reveal more information about the suitability of the products or services you plan on introducing to the needs of the target market.

You must approach market development systematically, or you might waste a lot of resources on a market whose size will be depleted soon after you spend on it.

Improve the Pricing Process

In any business venture, pricing stands out as one of the most important things you must focus on. Your pricing process will determine whether your target customers can afford your goods and services or not. Remember that one of the things expected of you as a Lean Six Sigma expert is to maintain customer satisfaction.

Exorbitant prices will chase most of your existing customers away and scare away those who might have been interested. Pricing strategies are also important to your bottom line. It is possible to improve profitability by up to 25 percent by reviewing your pricing strategies. Other companies have managed up to 60 percent. The following are useful ideas that you can combine with your Lean Six Sigma knowledge to improve your pricing strategies.

- *Value addition*

To improve your business margins, you must consider value addition. In value management, focus on what the customers want and the things that they hold dear. When selling to your customers, the message they need to hear from you is the value they are paying for, not the price they are paying for your products or services.

Customers are more inclined to appreciate the pricing strategy when they understand the value they are getting. More often, those who understand value will have knowledge of the market demographics, so they know what is offered elsewhere, but they choose to remain with your company because of what they get from you.

- *Shun price wars*

By default, price wars have never been won. Try and resist the urge to engage in a price war with the competition. Market forces eventually take precedence, and the price wars become unfruitful. Instead of engaging in a price war, understand what your competition is about.

Improve Advertising Communication

Customers consistently depend on advertising communication to learn more about what your company is offering. You can address advertising communication in the following ways, which will help you get them the information they want in a timely manner:

- *Sales promotion*

Sales promotions can be scheduled over the short-term or medium-term. During this process, give discounts, incentives, and price reductions so that you can kick-start sales in a given product. Sales promotion is also a good way to reinvigorate products that might have been sluggish in sales. You can also use sales promotion to wipe out dead stock from the shelves.

- *Personal selling*

The beauty of personal selling is that you have one-on-one interaction with the customer. This is a golden opportunity for you to introduce a service, promote, or demonstrate something. Since the customer can interact with you personally, you have a chance to woo them over and impress them. You can also get their feedback instantly. Companies these days use personal selling as an opportunity for customers to interact with some of the senior level management, which can boost their enthusiasm for your company.

- *Public relations*

Public relations might work, but you have to structure it appropriately to guarantee results. Since you have no control over what the media outlets you share the information with will do, you must strive to do a good job at it. Ensure the endorsement you expect of your product is impartial to maintain higher credibility to the customer.

- *Internet marketing*

Everyone is using the internet these days. If your company is yet to embrace internet marketing, you are shooting yourself in the foot. We have many tools available for internet marketing, which are effective in delivering the required results. You can use your website, social media channels and any other tools available at your disposal. You will need a good social media manager for this, one whose roles will be in line with your Lean Six Sigma strategies.

- *Direct marketing*

Direct marketing works in the same way as personal selling, just that you barely get the face-to-face. Instead, you use any communication methods at your disposal to contact the customer directly. You can call, send a letter or samples either to their home or office address. Direct marketing is usually a good idea especially when dealing with high profile customers because they receive the news before anyone else.

- *Advertising*

You can also pay for nonpersonal presentation to promote your products or services. While traditional media outlets like radio, TV, and billboards are very expensive, they also have a higher success rate. Today, however, businesses are venturing deeper into internet marketing, especially those who are creative but lack the financial outlay required of traditional media advertising.

Assessing Customer Satisfaction

Since Lean Six Sigma is about improving processes, it follows that by improving quality production, you should also find a way to improve sales. All companies have their wants and needs, just the same way humans do. These wants and needs must be arranged in a hierarchical manner so that you can understand the importance of each aspect. Hierarchy is important because it provides an outline of how to meet and exceed customer needs, based on the nature of your relations with them.

Borrowing on Maslow's need hierarchy, to satisfy an individual, you must first appeal to their most basic needs like the need for water and food. From there, you can go a notch higher and offer them things like esteem, love, and security. It is only when you satisfy someone's basic needs that you can empower them to self-actualization.

The Kano needs analysis provides a great tool for determining and addressing customer needs (Coleman, 2014). Your sales team should incorporate this into their work to give them a better impact in their approach to customer needs. This analysis gives you a good idea of the basic customer needs. According to this concept, a customer can only be satisfied to the extent to which the product or service you are offering provides them utility.

This analysis is built on three requirements. It must satisfy the following needs:

- *Basic needs.* These are the requirements that you use to get your company into the market and get the customer's attention. These are also referred to as dissatisfiers.
- *Performance needs.* These are the requirements that will allow you to keep operating and competing

favorably among your competitors even when economic conditions are not favorable. Performance needs are referred to as satisfiers.

- *Excitement needs.* Excitement is about growth and development. They allow your company to excel and grow by offering world-class services or products. Excitement needs are also known as delighters.

Dissatisfiers

These are features of your service or products that are expected. They are obvious. The customer does not need to ask for them. They are the bare minimums, and if you do not meet them, the customers will be disappointed and move on to another product.

Dissatisfiers are often unspoken. They are common sense requirements. A good example of an unspoken need is cleanliness when you go to a hotel. Cleanliness encompasses everything from the entrance, the hotel lobby, to the hotel room, the bathroom, linen, and the air.

These are basics that a customer does not need to ask for. When you book a hotel room online, you do not need to ask for the room to be clean, and you expect that it should be. If this is not the case, you will express your dissatisfaction, and take your business elsewhere.

Satisfiers

Performance needs are standard requirements that increase your level of satisfaction. Assuming that the customer's basic needs are met, the performance needs are a bonus. These include things like speed of service delivery, ease of using appliances, commendable prices and so forth.

Unlike dissatisfiers, performance needs can be discussed and spoken. Let's go back to the hotel example again. A customer will be very happy when you heed their request to move them from a room near the elevator to a quieter one. Do you have a smoking room? If they request and it is available, this will make them even happier.

Delighters

Delighters are surprise packages, unexpected features that earn your sales team additional points in favor from the customers. Naturally, delighters are barely spoken. When you stay at a hotel, they can surprise you with some of their customer-favorite cookies in your room, without additional charges. This is something the customer does not expect, but when it happens, they will be surprised and happy about it.

When implementing Lean Six Sigma, you must try to be as flawless as possible to meet customer needs. Dedicate all resources necessary to meet the customer needs, and you will be happy with your business performance.

In your capacity as a Lean Six Sigma expert, you should have the knowledge to understand the basic needs and acknowledge them, but not to dwell on them too much. They are rudimentary and dwelling on them might not auger well with the customer. Meet them, but move on to the higher level needs.

Experts advise focusing on delighters because they give the customer extraordinary benefits without necessarily increasing your cost overheads. In line with the Lean Six Sigma methodology, you have sufficient tools and resources at your disposal to understand, develop and deliver products and services which add value to the customer. Value addition increases the chance that the customer will be very happy and stay with you than move to your competition. A lot of businesses run short of time or money to spend on improvements. Because of this reason, following the Kano approach gives you a shortcut to determine what you should do and what you can avoid.

Over and above the three requirements, the Kano model also includes indifference and reverse requirements to meet customer needs.

Indifferent Needs

Indifference refers to features that will not have any effect on the customer's level of satisfaction or your performance, whether they are available or not. Still using the hotel example. The customer remains indifferent to the inventory management system that you are using. It does not concern them. The inventory management system does, however, matter to you because it affects the way you meet the customer needs. All the customer wants is to get to the hotel and be happy with the services they receive.

Reverse Needs

The presence of reverse needs gives the customers a negative perception of your business. They will leave a negative effect on customer satisfaction. The problem with reverse need is that at times they complicate the dissatisfiers, and customers barely get the chance to experience your products or services.

If they come to your hotel and find it so crowded that the queues are not moving and people are complaining about your inability to handle the crowd effectively, they will be disappointed and might opt to find a better option.

8

HOW TO IDENTIFY IMPROVEMENT PROJECTS

Not all processes or projects need improvement. In some cases, assigning unnecessary improvement tasks to projects that are already efficient could create waste in a process that was already efficient. Success with your Lean Six Sigma project is highly dependent on how well you can choose the right one. Choosing the wrong project can create unnecessary pressure and waste more time.

There are lots of negative consequences that you will suffer when you choose the wrong project. Some of these include:

- Getting the project canceled

- A higher cost of running the project
- Longer lead times
- Lack of motivation from team members
- Lack of attention and support from management
- Ineffective project results

In light of these challenges and risks, it is important that you try and take your time analyzing projects before you choose the right one. The following are some useful tips that will help you learn how to choose an appropriate project.

Consultation

Collaboration and consultation are some of the key tenets of Lean Six Sigma. Before you choose a project, make sure you consult widely. Reach out to your fellow employees or the main project drivers and find out the problems that they are dealing with. Learn from them about the challenges they face in meeting their deliverables—the things that impede their performance—and use those as the first things you can improve.

The good thing about consultation and collaboration is that you empower employees, and they feel they are part of the resolution. This is a motivator that will work in your favor down the line. Perhaps the biggest concern about this project is not being able to understand the sheer magnitude of the problem. Is it something that only a few people are experiencing or is it a major issue?

After consultation, you must use your Lean Six Sigma experience to categorize the problems that your employees have mentioned as SMART problem definitions. A few concerns might peak your mind, and you would have to look into include improving timeliness and accuracy, reducing the workload, or reducing the direct impact that repetitive administrative tasks have on the employees.

Using Performance Indicators

The use of performance indicators is a structured approach that is driven by process needs. You will focus on process indicators like claims, incidents, irritations, and lead times. Look at the projects available, and identify those that show consistently poor performance. The idea here is to ensure that you can analyze the indicators to reveal a good alignment between the projects and their place in the priority list.

While this procedure will be effective, you risk sidelining your employees. They feel downcast and sidelined from the decision-making process involved in choosing the appropriate projects. Apart from that, there is no guarantee that you will always have all the indicators needed. Some of the indicators might be available but unable to provide reliable metrics.

Using Performance Reports

Another way of choosing the right project for Lean Six Sigma is to use performance reports from previous months or periods. Analyze these reports to identify the areas where the company is experiencing major deviations from the budget or your plan. Some of the common areas include investment and IT maintenance and personnel overheads. These are notorious sectors where the overheads are often higher than the budget allocation.

This approach is useful to you because you will end up with projects which focus on improving results over time, and the results are measurable. You will see the result of your Lean Six Sigma approach in the subsequent monthly reports.

Benchmarking

Benchmarking pits your company against the competition or industry standards. Compare your financial or operational performance with the industry expectations, especially if you have access to that. Your Lean Six Sigma project should begin in the projects or processes where you feel you are underperforming.

While this approach might be useful, it has some challenges. In most cases, it is not easy to identify the ideal benchmark, and even if that were available, you might have many reasons they might not be applicable in your organization.

Strategic Decision Making

The most effective way to identify Lean Six Sigma projects is to look at your strategic position as a company. Study the mission, vision, and long-term goals of your organization. In case your department or division is to be held accountable in any of these, you can take responsibility for that and make it your Lean Six Sigma project.

The good thing about being a project head in this manner is that you get to choose how the goals or success in your endeavors are measured, and how to express them. Make sure that you can express them in a measurable manner.

Based on previous performance, you can set a target to improve customer satisfaction from 4.6 to 6.0 during the current trading period. Once you meet these goals, you can then take a step back and look at the short-term goals and how you can use them to achieve your overall goals in the long term.

Using short-term goals as a guide can help you identify the challenges that you might have to overcome if you are to meet these goals and project these into long-term targets. Using the same example of improving customer satisfaction, if you were to achieve a 12 percent increase in five years, you should first try to get a 4 percent increase within the current trading period, then build on that.

While strategic decision making is a success in identifying Lean Six Sigma projects, you will have to sit through a lot of meetings to determine the short and long-term goals according to the plans, mission, and vision of your company. This can at times be counterproductive, especially when some employees fail to see the need for the meetings.

Project Champions and Master Black Belts

A strategic plan is mandatory to meet the needs of the company. Everyone who is involved in the process should have a good understanding of the business process. You must understand what is driving the business and what is keeping the business together. When you look at all the conflicts that happen in the company from time to time, how is it that the business still manages to stay put and competitive in the market?

What holds the company together is the framework upon which they collect input such as resources available and consumer requests. This framework also enables the company to determine a positive outcome. Anyone who understands the business framework understands the core of the business. This is the same methodology that is used to determine the processes that are redundant and the ones that need improvement. It also helps to make sure that the projects chosen are aligned with the overall action plan according to the company strategic plans.

The people in the company whose role is determining the Lean Six Sigma projects should be properly trained in Lean Six Sigma. These are usually master black belts and project champions. When working together, master black belts and project champions analyze business plans to find out the core areas of the business process. This is where project selection starts. These are also people who have the best knowledge of the overall goals and objectives of the business, and how they are tied to Lean Six Sigma. They also understand the contribution to overall business processes by individual departments, and from their experience, they can easily tell how each business department or process contributes to the overall goals of the company.

The role of project champions is to collect and arrange information for master black belts to use on the selected projects. Some of the information under question include how the interconnectivity between departmental output, concerns regarding staffing, funding constraints, challenges with time management, and so forth. They also analyze the internal and external factors that are critical to meeting business goals.

On their part, the master black belts are tasked with assessing all the information presented by the project champions, and setting a hierarchy for deploying projects according to strategic priorities. They will then get the project running, supervise and assign tasks and roles in different aspects of the project. In some cases, the master black belt will be called upon to train and mentor other employees.

9

HOW TO WIN MANAGEMENT SUPPORT

A lot of projects fail because they never get support from management. The same applies to Lean Six Sigma. It is impossible to implement the strategies and succeed without getting the management on board. One of the challenges you might experience in this is dealing with management who do not understand Lean Six Sigma or why it is important.

Together with your team, you will discuss and outline the information you must share with management to get them on board. You need the company to provide the resources necessary, and this is one of the reasons you must get management on board. Apart from that, having the management support is a show of commitment that will reverberate through the company, and motivate the employees to work harder.

Importance of Management Understanding and Support

The more knowledge the management has about what you are doing, the easier it is for them to understand what you need from them and why. Implementing Lean Six Sigma methodologies is not always a walk in the park. You will encounter resistance from time to time, especially because everyone is not committed or welcome to change.

Maintain the Company Culture

The first thing you should always understand is that the business goals and Lean Six Sigma goals must always coexist. If at any given time these are pulling in opposite directions, you will fail. Your work as a Lean Six Sigma expert is to blend your task goals with the strategic objectives of the company.

Your position as a master black belt might delude you into thinking you have the overall say into company matters. However, the management team is the one tasked with integrating your objectives and goals into the company culture. This is why they should never be sidelined.

Lean Six Sigma is not about having one employee working on something in isolation to improve the business processes, and it is something that is contagious and has consequences. There will be frustrations in equal measure as there will be milestones achieved. Implementing Lean Six Sigma in any organization goes over and above any department, team, or employee. You, therefore, must make sure you have the management on board to oversee business improvement alongside your efforts.

Leadership Paradigm Shift

As a leader, you must understand where you cannot exert your influence beyond a certain limit. This is where you bring in the management. You must work together to succeed in implementing Lean Six Sigma. Try and make sure that you do not give off an impression of hunger for power. Most management teams do not take kindly to this.

Instead of individual project improvement, your work is to ensure maximum quality improvement. Lean Six Sigma experts especially those who have attained master black belt status are trained to be facilitators and contributors.

You should bring in the management as facilitators so that they play their role in driving change. This would be a commendable approach to assure the company that you are not overreaching in your roles. As facilitators, management is tasked with providing support, encouraging the rest of the workforce to cooperate with your efforts, and participate fully in the processes recommended.

They will also motivate the employees and rally support for you, showing their commitment to your deliverables, supporting you and laying the foundation for future success. If you come on board as a consultant, you should prime the management in their roles as facilitators in case they are not aware of what is required of them.

Their ability to support the employees and champion them to work toward success is more important than the knowledge of Lean Six Sigma approaches. However, you should still try and encourage them to learn.

If management is aware of how to handle the methodologies

you recommend, it is easier for them to push their employees to get on board and deliver the best of their abilities. This way, you will have incorporated the business goals into the Lean Six Sigma requirements and will be on your way to success.

As a consultant, or if your company brings on an external Lean Six Sigma expert, you must prepare the management ahead of everyone else. You must give them specific training that will get them ready to support your cause. The start-up point is often the most critical part, and this is where you get the management on board.

Training and Education

Unless in rare scenarios, management teams are often leaders and experts in what they do. These are people who have mastered their roles for years. There are many skills that they have which will overlap with what you are bringing on board as a Lean Six Sigma expert. However, since you might not be lucky to find a management board that is adept in Lean Six Sigma, you should be ready to offer the necessary training to bring the best out of each of the managers. Once they grasp the basics, they should be able to proceed without any challenges.

Structure the training in a way that can help management develop projects which support their strategic objectives. If resources are available, schedule training sessions for the entire company, alongside session per department. Reach out to the finance team in your company to ensure they can support your requirements within their allocated budget.

Overcoming Management Resistance

You might get support from management to implement Lean Six Sigma, but not all the time or everywhere. You can come across instances where management is reluctant to support you, or if they do, they do not go all the way. The following are useful points that will help you get management to reconsider and support your Lean Six Sigma cause:

- *Result assurance*

You have to show management that you can realize the project goals and benefits in record time and help them meet their cost considerations. On average, you should strive for 30 percent return within five weeks. This is a proposal that will get most management teams to consider your offer. Once you have their attention, show them the possibility of projects that are self-funded, which means they will only need to offer an initial allocation and let the project run its course. Even as you make these promises, ensure they are realistic and attainable.

- *Astute project selection*

How do you choose projects? The criteria you use should show management where your priorities lie. If they see you are on the same page with their strategic plans, it is easier to support you. The projects should have a clear outline, which means they can tell how soon it will take to complete them. They also need to see the tangible benefits of supporting you

- *Setting departmental goals*

Encourage your teams to discuss and set their goals. In practice, people are more motivated when they set their goals. This is because they try to avoid embarrassment. Motivation is important for improvement projects because everyone will be looking toward completing the projects on time and to see the results and how they blend into the company goals.

- *Incorporating Lean Six Sigma*

In the presence of a capable team that is well-versed on Lean Six Sigma, it is easier for them to deliver the small wins you need to impress management. When they notice a reduction in variation and cycle times, management can offer you the support you need.

- *Progress monitoring*

Your project must be aligned around a clear work plan. You must show milestones and when they will be met. These provide a blueprint within which management will try to understand how your Lean Six Sigma plan fits into their strategic goals. With a clear work plan, it is easy to keep tabs on progress and identify the strengths or challenges of the recommendations you are implementing.

- *Support people processes*

By default, most management teams are more welcoming to people processes than those that encourage chemical reaction or machine processes. They will be encouraged by processes that provide room for eliminating waste and redundancies, standardized work, and removing activities that add no value to the team.

- *Team dynamics*

If you need management to change their stand and support you, you must also show them that your process is team oriented. The success of your project will, in most cases, depend on the quality of the team you are working with. The Lean Six Sigma project leaders must ensure the team is coherent, and there is a healthy rapport.

COMMON LEAN SIX SIGMA MANAGEMENT IMPLEMENTATION MISTAKES

P eople make mistakes all the time. What matters is the lessons you learn from the mistakes, and whether you use the experiences as a platform to improve your life. For any project in management, mistakes are bound to happen. This is why you should always make plans and allow room for rectifying such, especially when presenting expected deliverables and deadlines.

Some mistakes can be catastrophic. In Lean Six Sigma, you are learning how to improve efficiency in your workplace as a manager and eliminate redundancies. Even in this process, mistakes might happen. Professionals have made mistakes in the past which have been well documented, and you can use these as a learning point. Learning about common mistakes people make in Lean Six Sigma is a way to get you prepared so that you can tell the signs in good time and prevent yourself from catastrophic challenges.

Impassive Leadership

Apathy in management can set your company back so many years. To effectively implement Lean Six Sigma, you need a strong management team that will not just motivate the workforce but will also provide the necessary leadership.

One of the common challenges that you experience in the struggle to implement Lean Six Sigma is having a management team that is indecisive and disinterested in what you are trying to achieve. The more resources you spend to improve the team, the easier it will be for you to succeed with the same team.

What you need is full support from the leadership. To get this support, highlight the importance of learning and implementing Lean Six Sigma, in the process setting a good example for the rest of the team. You need the team to be motivated and effective in what they do, and this motivation must start from the top brass.

Inapposite Deployment Policies

The policies you initiate for your team will determine whether you succeed or not. From the blueprints, you must ensure your policies are geared toward success. To do this, analyze and evaluate your strategies frequently.

Careful analysis should reveal the depth to which the company problems run, and what can be done to improve the current position. Many times managers come up with succinct policies in a bid to implement Lean Six Sigma in the workplace, but the policies do not work because they were not carefully structured for the company or the industry.

Avoid one-size-fits-all policies. Policies, strategies, and plans that you prepare for your company should be backed by accurate data. Consider all possible variables to increase the probability of success.

Extraneous Compulsion for Advanced Training

Training is important to improve the performance of the company. However, you should be careful about pushing your team to get their certification. Ensure you have a good reason to champion training. Without valid reasoning, you might end up spending too much on training with very little to show for the same.

A good example, you might end up having everyone attain their yellow belt certification but lack the guidance of higher tier Lean Six Sigma levels in the organization. As a result, you will only have improved performances in one level of the organization, stagnating growth in all the others.

It is advisable that you exercise caution in the manner you go about training so that you can apportion roles and tasks accordingly, depending on the certification that your employees have already attained.

Using Training as a Divertissement

In many organizations, people do not take training seriously. When there is undue pressure for indoctrination, at times the entire process becomes redundant and distracting. People end up looking forward to the training sessions just so they can get out of work. Over time, they lose focus on what is important.

If you keep peppering your employees with Lean Six Sigma

training all the time, they will barely focus on the basics. Losing grasp of the basics is the first step toward failure. Instead of focusing on training, you should invest more in coaching and mentorship programs, to ensure that those who need support get it. Besides, everyone has a different learning curve, so you cannot expect everyone in the team to learn at the same pace.

Inappropriate Project Selection

Choosing the wrong project is just as bad as handling it poorly. A mistake most people make even with the best Lean Six Sigma techniques is poor handling of their projects. Once you lack focus and cannot prioritize project deliverables, you will barely get anything right.

Projects should be assigned depending on the expertise of the people leading them. Taking a project that is not within your capacity can easily end up in disaster. A yellow belt professional, for example, might not have the expertise and experience required of a black belt, hence struggle with black belt level projects.

Disparate Distribution of Effort

For you to succeed with Lean Six Sigma, you must collaborate with others in the team. Each person should play their role as you collectively work toward the ultimate goal. If the team members are not working together toward the same goal, there is a good chance you will struggle to achieve your goals, and all the Lean Six Sigma techniques will have gone to waste.

It is not just about understanding their roles, but you must also ensure they understand the benefits the team gets when they work together. Your role as a team leader should be to make sure everyone knows how interlinked their roles are, and to align their personal goals with the business goals.

Inaccessible Resources

A project is always going to be a team effort. However, resources should only be handled by champions. All resources the champions need should be available to them. The lack of resources like skilled team members in certain levels can derail the timelines, scope of the project, and eventually your success. Before taking on a project, therefore, make sure you get an inventory of what is required.

Lack of Purview

Before you take on any Lean Six Sigma project, make sure you know how broad it runs. If you do not look at the scope of the project, you might end up suffocating everyone in the project, struggling to punch above their weight. Look at the project outline. Make sure you know the objectives, the challenges involved, and assess whether you can address them accordingly.

Stakeholder Support

You might get the support of the management for your project, but the main stakeholders do not support it. This position poses a challenge to the management in that your project is not a priority for them. Given the interaction that other team members might have with some stakeholders outside of the workplace, they might be privy to some information about the dislike of the project or some reason for hesitation. As a result, there will be no motivation on their part.

Once you have support for your project from the stakeholders, you will be comfortable exercising your roles. Other than that, you will also have the resources you need to work on the project and achieve your goals.

BENEFITS OF LEAN SIX SIGMA METHODOLOGY AND CULTURE

Whenyou enroll for a Lean Six Sigma certification, you are combining two of the top management practices, Lean and Six Sigma. Both of these techniques borrow from each other, but they are not the same. Along the course, you will learn the differences between the two. In particular, Lean places emphasis on waste elimination, tasks that are not value-added, and variation. Lean is about commitment to creating a system that focuses on customers and improving service delivery by optimizing the flow of materials and information.

On the other hand, Sigma should be undertaken once you have implemented a commitment to Lean and established the process in your organization. Sigma takes a data approach to problem-solving. With data, you are in a better position to understand the variables you are dealing with and how they affect the process. This understanding helps you map a better way to optimize processes and solve problems.

Why should you consider enrolling for a Lean Six Sigma course? What benefits do you stand to gain in the long run?

Time Management

One of the biggest challenges that most companies have is the struggle with time management. Time is wasted in so many organizations, even the big global brands. Through Lean Six Sigma, you learn useful ways of making use of the time you have. You need to ensure your workers are productive and utilize the resources allocated to them efficiently.

When setting goals, one of the key concepts you should never ignore is time management. First, you set SMART goals, then incorporate the Lean Six Sigma principles into your plan to supercharge your goals. Three elements that you must introduce into the plans are fulfillment, performance, and learning.

You can study yours or the performance of your employees and determine how much time you take away from work to handle disruptions. You should also ask yourself how many of those disruptions need your attention or whether someone else can handle them.

Circumvent the Cost of Failure

Most, if not all, of the top companies in the world have implemented Lean Six Sigma into their management process. One of the key benefits they derive from this is millions of dollars in savings. By streamlining communication and internal processes, companies can avoid spending on unnecessary processes or resource wastage through ineffective communication practices.

The benefits regarding the cost of failure are not limited to business entities only. A lot of municipal governments and government agencies have also realized the benefits of Lean Six Sigma. Lean Six Sigma is a champion for optimization. As long as the entity runs its cost centers through a Lean Six Sigma approach, they establish a commitment to proper service delivery all through the organization.

Through the implementation of Lean Six Sigma strategies, entities have realized they can increase efficiency in performance. One of the other elements that are associated with Lean Six Sigma is teamwork. You instill a culture of team spirit, and everyone who is part of the process gets empowered to own their roles in the project or process, even if it is not directly associated with their daily roles in the organization.

To realize these advantages, you must ensure that all projects include people who are highly trained in Lean Six Sigma, especially those who have attained the highest levels of Lean Six Sigma training. These experts bring to the project their experience in project team management, development, planning, organizational leadership, analyzing measurement systems, and effective communication strategies.

In your role as a Lean Six Sigma expert, your certification and training makes you an important asset to the company, especially since you will often be called upon to identify and evaluate processes that need improvement. Over time, the entire company benefits from a spillover effect, through continuous efficiency in management and use of time and other resources.

Supply Chain Management

Lean Six Sigma has many benefits for the company. One of the goals behind this methodology is to ensure that your processes have the lowest possible rate of defects. To achieve this, you must involve the suppliers. A possible way out of this would be to reduce the number of suppliers. The lower the number of suppliers you have to deal with, the lower your risk of defects will be (Martin, 2014).

When you are implementing Lean Six Sigma, you must also consider the impact that this might have on your suppliers or their decisions on your business operations. Any changes that the supplier makes will certainly affect your business in some way. To succeed, you must try to make Lean Six Sigma improvements all through the supply chain.

Customer Loyalty

One of the core goals of a business is to keep their customers. Successful businesses have their customers coming back for their business all the time. To keep customers loyal over time, you must ensure the highest level of customer satisfaction in your business processes.

Most customers complain about not staying loyal to businesses because of the attitude of the employees and their experiences interacting with the company. The problem with customer dissatisfaction is that the company will barely be aware unless the customer lodges a formal complaint. However, they will simply move their business wherever they feel they are appreciated. Lean Six Sigma provides the tools you need to minimize the risk of losing customers to your rivals.

Strategic Planning

Strategic vision and planning is supported by Lean Six Sigma. In your work plans, you must first ascertain that the business has a clearly outlined mission statement. With that in mind, you can also perform a SWOT analysis or any other analysis necessary. With the results at hand, you have enough room to implement improvement recommendations through Lean Six Sigma.

Assuming your company's strongest point in the market has been your position as a market or cost leader, you can use Lean Six Sigma procedures to fine-tune your internal processes. This way, you can do away with complexities, improve your processes, and increase your expected yield. Another secret would be to try and bargain or maintain manageable supplier agreements. Whichever strategy your company uses, Lean Six Sigma will help you get the best out of it.

Improving Efficiency

One of the main reasons undertaking a Lean Six Sigma course makes you a good fit for any company is the desire to improve efficiency. Improving efficiency is something that most companies work toward at all levels. When you attain black belt certification, you are trained in analyzing processes in a workflow to determine where weaknesses might be present and propose a means of eliminating them. Eliminating these weaknesses increases the prospect of efficiency through resource and time allocation.

Achieving improved efficiency can be done either as a whole or in segments. If the company is segmented in a way that each system functions as an entity, each of the entities will benefit from improved efficiency, in the process benefitting the entire company as a whole. It does not matter whether you are working for a small or a large company, for your Lean Six Sigma expertise will always come in handy where efficiency is the subject matter.

Motivate Your Employees

You need your employees on board if you are to succeed in whatever you are doing. However, they need to be motivated to give you what you want. Most corporate entities that keep their employees motivated enjoy productivity boosts of between 25 percent and 50 percent. Motivation is not as difficult as most people think. At times all you need to do is engage your team, and make them feel they are part of the process. Through Lean Six Sigma you will come across useful tools that can help you come up with the right business environment that can spur employee motivation.

Output Quality

Efficiency and output quality go hand in hand. Lean Six Sigma focuses on delivering the highest possible quality levels. To do this, you need to place an emphasis on important operational sectors in the company, such as inventory control, quality control, product scheduling, and quality assurance. To achieve the desired output quality, you should aim to make sure quality concerns are nonexistent in all the operation processes.

Your first step would be to identify the set standards of operation in the processes. From there, you can recommend a practical approach through which different members of staff can be empowered to improve and own up to the improvements implemented.

Improved Customer Services

By improving the efficiency in the operations, it follows that the company will also improve their customer services.

However, this does not just happen overnight. Improving customer services is a compounded effort that comes from accurate production schedules and reducing the time needed to meet customer demands. The sooner customer needs can be met, the easier it will be to make the company's customers happy.

Workplace Safety

Workplace safety is an aggregate of many factors. It is not necessarily a direct benefit of undertaking Lean Six Sigma certification but a bonus. Everyone enjoys working in an enabling environment where things run smoothly. Workplace conflicts will happen from time to time, but this should not be a problem because you should have procedures and steps to address these as and when they happen.

Lean Six Sigma inculcates a culture of commitment to the workplace, and employees own their actions. Therefore, in the long run, Lean Six Sigma champions a safe work environment where everyone understands the corporate organizational culture and their roles in their respective departments.

Manage Deadlines

A lot of projects tend to run beyond the intended deadlines. A careful inquest reveals changes in management policies or change in the project requirements or scope, among other things. When you consistently fail to meet deadlines, it becomes a problem that can manifest and become a system of inefficiency.

Inefficiency should never be allowed to become the norm. In Lean Six Sigma, you will learn how to structure your team so that you have experienced personnel at different strategic positions in the business or at the necessary departments where they champion the changes you want. One of the key roles of those in these strategic roles is to make sure that they can identify some of the factors that prevent you from meeting the deadlines.

Other than identifying the problems, you can also use the same personnel to identify solutions for the problems. In the long run, you succeed in creating shortened project cycle times. As long as you can maintain and meet the deadlines, you can reduce cycle times by around 35 percent.

Project Implementation and Management

When a Lean Six Sigma expert joins the team, more often they change the way projects are implemented. In some cases, this can introduce a culture shift from the way the company is run. Therefore, before committing to Lean Six Sigma, you must understand the changes that the methodologies will bring.

To improve processes through Lean Six Sigma, you need to ensure that the dynamism is reviewed objectively. The Lean Six Sigma structure involves process creation, process improvement, and process management. For the company and individuals involved, it is important to look at Lean Six Sigma as a concept, not an assignment. The Lean Six Sigma ideology should be embraced throughout the company if they are to succeed.

Profit Motive

Businesses are finding it harder to manage the cost of operation with each passing financial year. Through Lean Six Sigma, you can learn how to deal with the rising costs of operation and the expected threat from competitors. Your role will be streamlining the business operational processes, in the process increasing the company's profits.

The benefit of streamlining operational processes is that you end up with faster task completion and improved efficiency without compromising the quality. In retrospect, Lean Six Sigma is about increasing expected revenues by allowing the company to achieve more through full utilization of the existing resources.

Cost Consideration

Having looked at the profit motive behind Lean Six Sigma, it follows that there must be a considerable reduction in operational costs. How does Lean Six Sigma make this possible? The core of Lean Six Sigma is about eliminating waste from all processes. How do you identify waste? Most people see waste as a by-product in the production process, something that is discarded. However, waste can also become part of the system.

In Lean Six Sigma, waste is an activity that exists within the process but serves no purpose in as far as the production process or service is concerned. You will also need to solve problems that occur as a result of certain processes. When problems exist in service provision or production, they become additional costs to the company, which should be eliminated. Lean Six Sigma, therefore, helps you manage overheads by making sure the company gets utmost utility from the resources available.

Team Spirit

A successful company must thrive around strong interaction between teams and individuals within the company. Not all employees in their ranks are effective. To change this, one of the lessons you will learn in Lean Six Sigma is how to involve the workers in the process of improving service provision or production processes. This is aimed at building rapport and encouraging active participation from all employees.

By including employees in the decision-making processes,

you are also creating a culture of trust and transparency. Employees perform better when they know their input is appreciated, and they feel appreciated too. The trust gained is important because everyone believes they have a role to play in the organization, and they are just as important to the overall success as those at the top level management positions.

Easy Learning Process

There are several programs that you can undertake in a bid to improve quality management in the organization. However, of all the programs, Lean Six Sigma is one of the easiest to go through. One reason this is true is that Lean Six Sigma is segmented into stages. Each of the learning stages is accorded a specific mastery experience.

The lessons you learn in Lean Six Sigma are applicable in many fields, including corporate ventures, service, and manufacturing industries. This is one of the most flexible quality management courses that you can undertake, and the best thing about it is that the lessons you learn are applicable in many business ventures, irrespective of their size.

How to Enjoy the Benefits of Lean Six Sigma

Management must have a good understanding of the intrinsic value of Lean Six Sigma for them to appreciate the gains it brings to the organization. Contrary to the opinion that Lean Six Sigma is employed simply to save money, it is an all-around process that, when implemented accordingly, stands to favor all departments in the organization.

The key to succeeding with Lean Six Sigma is astute governance and discipline. For Lean Six Sigma to be effective, the results must be tangible and capable of validation from the respective departments. You should remember at the back of your mind that this is not a one-time thing. Lean Six Sigma requires patience. It might take time, but upon successful implementation, you will appreciate the effort and time spent. To succeed with implementing Lean Six Sigma, the following are some of the prerequisites that you must have as a company:

- *Needs assessment.* Your company must have a very good reason behind implementing Lean Six Sigma. You cannot invest in this process just because you have money to throw at it. With a compelling reason, you can take stock a few years down the line, to see whether your investment into Lean Six Sigma was a success.
- *Management support.* You need the full backing of senior management if you are to implement Lean Six Sigma. They must be dedicated to the cause, and support the process by all means possible.
- *Resource allocation.* The company must also set aside sufficient resources to ensure the appropriate training is available. Resources vary and can include anything from personnel, technologies, and materials.
- *Common goal.* All the stakeholders in the company must come together and work toward achieving the overall objectives of the company.
- *Responsibility.* All team players involved in the different Lean Six Sigma processes should be encouraged to take responsibilities for their actions. This is a process that eliminates the need for constant supervision, approvals, and evaluation, which consume precious time.
- *Qualitative analysis.* As the company maps out the priorities behind the decision to bring in Lean Six Sigma, you must make sure you focus on making changes that will introduce quality into the company

over the long term, over changes that bring instant or immediate results.

- *Innovation.* One of the simplest ways of encouraging a culture of innovation in the company is to make sure your personnel are not spending a lot of time struggling with problem solutions or dealing with challenges. The same applies to management. Lean Six Sigma helps you streamline procedures to the point where your team has more time on their hands for creative thinking.
- *Strategic positioning.* You will notice that operations in Lean Six Sigma require flexibility in response to dynamic business environments. Therefore, you have to adapt so fast in response to unplanned changes either in the industry or the economy.
- *Set and maintain standards.* Lean Six Sigma strives to eliminate variation. In so doing, standardization is encouraged. Other than standardization, this process also encourages project management, training personnel, and monitoring employee performances. These procedures help to streamline and simplify the corporate operations.
- *Competitive desire.* Lean Six Sigma professionals often exude a competitive aura about them. The desire for excellence and success is something that will make you competitive in most, if not all areas of your business operation.

LEAN SIX SIGMA CERTIFICATION

Certification is important as you make plans to begin your managerial career. Getting certified opens doors for you, to companies and industries that will create new opportunities for your career. As a professional, certification helps you make moves in the job market that you will appreciate years down the line. You have better chances at interviews and better opportunities than your peers who are not certified.

Through certification, you will revisit the key principles of

Lean Six Sigma and learn how to align your personal and career goals with the objectives of Lean Six Sigma. The basic understanding of Lean Six Sigma is a systemic and rigorous process where you improve operational performance by using tools in your possession and, at the same time, identifying and getting rid of defects in your place of work.

Lean Six Sigma is applicable directly in almost all manufacturing plants and processes. By getting your Lean Six Sigma certification, you are equipping yourself with the necessary tools and skills that will help you identify bottlenecks within your work environment. The skills you learn are not restricted to the manufacturing industry alone. By the time you qualify and get Lean Six Sigma certification, you should also know how to improve performance in different situations within the service industry.

Why is it important that you get your Lean Six Sigma certification? There are many other quality management and improvement methodologies in the industry. What sets Lean Six Sigma apart from them? Lean Six Sigma espouses the following three aspects:

First, it addresses the importance of and the need for decision making. Decision making in Lean Six Sigma is built on data that has been tested and verified through statistical approaches. In so doing, decision making is done through an evidence-based approach, not assumptions from a subjective point of view.

Second, Lean Six Sigma is an approach that is strongly supported by the echelons of leadership and management in whichever institution you work in. By enrolling for a certification, you are not enrolling into a one-off experience, but something that will transform your life. It is a cultural element that will guide your management career.

Third, Lean Six Sigma emphasizes quantification and measuring processes. It would be easier to evaluate and estimate the financial returns. Through Lean Six Sigma, you will make quality improvement projects tangible because they can be tracked, and from the results, assessments can be made concerning the goals and costs associated with them.

Lean Six Sigma Certification Categories

In a professional capacity, you need to take responsibility, own the quality processes that you handle in management. A Six Sigma expert is a professional who is an expert at improving processes to help the company achieve their target organizational goals. You must understand all the details in each step of processes you are involved in, how they work, and what can be improved. Allow yourself enough time to study and understand how the concepts and tools available can be used to improve the processes and turn them into a success.

Your role as a quality management expert is to guide employees under your wing to continually improve in their roles. In light of this, one of the skills you must possess is mentorship. Together with your technical ability, those assigned under your supervision should notice the improvements in processes and the way they accomplish tasks under your leadership.

To acquire these skills, you need Lean Six Sigma certification. There are four levels of certification in Lean Six Sigma, all classified as belts. The concept of belts is borrowed from martial arts. With each belt level you achieve, you are expected to have achieved a certain level of skill, meet set expectations, and manage certain responsibilities that are bestowed upon you.

White Belt

You might come across white belt workshops for Lean Six Sigma. The concept behind these workshops is to introduce you to the concept of Lean Six Sigma, and the fundamentals of what you are getting into. White belt workshops are about preparedness. You are introduced to variability, process improvement, and process performance, and you also learn the importance of each team member and their roles in your team.

As a student, you learn how to become an active team member or a project leader at the end of your certification. White belt students will also gain an introductory under-standing and knowledge of Lean Six Sigma, the DMAIC process and roles and how to reduce resource wastage while at the same time ensuring customer satisfaction.

Benefits of White Belt Training

If somewhere along your career you feel you are ready for a change, one of the positive steps you can make toward achieving your goals is to enroll in a white belt training workshop. This workshop is perfect for someone who does not have basic knowledge of management methodologies but has a desire to make wholesome changes in their career. The following are reasons white belt training will suit your goals.

- *Best introduction*

When you begin your journey into Lean Six Sigma without prior knowledge of management, you might have a difficult experience adjusting. A white belt workshop takes care of that problem. The white belt training helps you prepare for what lies ahead.

- *Time-conscious*

Like most people, you are worried about not having suffi-cient time to take on a course and balance the rest of your life. With this mentality, you might struggle with the rest of the Lean Six Sigma courses. What white belt training does, is it gives you a subtle introduction into part-time studies. It is a very short course, which can be completed in six to eight hours.

- *Laying foundation*

White belt training gives you the best foundation upon which the rest of your Lean Six Sigma training will be built. You learn how to use variance and the importance of role specificity in your projects.

- *Important certification*

White belt training is an important certification to have. With this, you should be able to join different teams or projects either as a team leader or an integral member of the team goals, depending on what they set out to achieve.

- *Beat the competition*

Professional certifications in management set you a cut above the rest. You get priority over the competition in most professional positions you will apply to.

- *Introducing DMAIC*

DMAIC forms the basis of Lean Six Sigma. It is one of the most important things you must learn. The five-step process will be used in improving processes almost every other day you are at work. At the white belt certification level, you are introduced to DMAIC and how it will influence the rest of your career.

- *Comfortable learning*

Since you are taking a course in your spare time, white belt training is offered in your spare time. You can attend any workshop available or get an online primer course so you can learn at your own pace and time.

- *Understand division of labor*

Division of labor is an important part of management. You need to know how to assign the right people to the right roles based on their competence. In white belt training, you will also learn how to assign roles to professionals and the kind of placement positions that suit your personality.

- *Value addition*

The best thing about white belt training or any certification you are interested in is value addition. By undertaking this course, you commit to becoming a better version of yourself, and for your career growth, you become an invaluable part of the organization.

Yellow Belt

Yellow belt training builds on what you learned in white belt training. It is about the fundamental methods that you use in finding solutions to problems. You will learn about basic procedures for improvement and the metrics applicable. Most people who have yellow belt certification are considered experts at specific subject matters.

At the end of this course, therefore, you should become a key member of a project team, either on a single project or more. The ultimate objective for yellow belt training is to empower you with the necessary tools to help you achieve the overall goals and objectives of your company and, at a personal level, to become an invaluable team player.

Benefits of Yellow Belt Training

The following are some of the advantages that you stand to enjoy when you enroll yourself or your team members into yellow belt training.

- *Team solutions*

With yellow belt training experience, you can champion solutions for problems at team level. We often end up in a scenario where some problems cannot be solved because of a lack of a structured approach. Yellow belt training teaches you how to involve the entire team in problem-solving, in the process improving the morale of the members because they become a part of the solution, hence feel a stronger connection to the team.

- *Project success*

Yellow belt fellows offer support to leadership in black belt and green belt levels. This tiered support approach makes it easier to achieve the overall goals of a project. Everyone in the project has a better understanding of their roles.

- *Time concerns*

You can complete a yellow belt training in two to three days, so you do not have to take a lot of time from work. You will learn the basics in this short duration, but the impact in your career will be worth so much. Besides, this is a good stepping stone upon which you can progress to green belt and black belt certifications.

- *Owning success*

Over time, you will learn the importance of owning your success stories. With yellow belt training, you get more people within the organization to identify with the improvement process that you initiate. Over time, you establish a system of improvement that becomes a tradition in the workplace and inducted into your cultural heritage as a company. Through yellow belt training, therefore, you can train, in a shorter time, so many people working under your supervision.

- *Compounding improvements*

Yellow belt training is about making the most of marginal gains. Making small improvements in different sectors of the organization eventually compound into massive gains. The beauty of yellow belt training is that you exploit the benefits of economies of scale. Yellow belt experts are known to induce small improvements in the right places. Yellow belt experts are the best embodiment of the power of ants in an ant colony. They are the backbone of transformational attitude in a business.

Green Belt

The green belt is considered one of the basic level of certifications for Lean Six Sigma. At this level, you are expected to understand the tools that are required in the Measure and Analyze aspect of Lean Six Sigma.

More often, your company could identify you among other members of staff for this level of training. After this training, you are then assigned to different roles in the company, often in charge of teams, while still keeping up with your normal responsibilities in the workplace.

Is green belt training suitable for you? It is perfect for entry-level positions, especially someone who needs functional, foundation knowledge on Lean Six Sigma. With this certification, you can handle small projects, especially those that require planning, action, and evaluation. For any project, you will need black belt and green belt professionals. However, green belt professionals are also needed to prop the roles of their superiors.

Green belt experts are taught about the essentials of using DMAIC in managing and improving projects that they handle. You also learn how to identify and eradicate problems by implementing suitable solutions.

Green belt training is also about people management. The kind of training that you receive equips you with the knowledge to oversee the commitment, expertise, and persistence of the members of your group. This plays an important role in improving processes. One of your roles as a green belt professional will be to provide a link between the real world application and the theoretical concepts learned in Lean Six Sigma.

Benefits of Green Belt Trainings

As a green belt certified professional, these are some of the advantages that you will soon enjoy as soon as you are through with the classes:

- *Financial growth*

Before you begin Lean Six Sigma classes, you might have looked at the financial benefits that come at the end of the course. As a certified expert, you are in a better position. You can now be trusted to manage projects of a larger value than what you might be used to.

Other than high-value projects, you also look forward to higher earnings. On the part of your company, they also stand to gain a lot from your expertise, making significant savings through your project management skills.

- *Strategic positioning*

Green belt training also sets you on a good path regarding strategic positioning. This certification upgrades your strategic role in the company. You should now be able to handle projects that have strategic significance to the company, especially when it comes to problem-solving. You are also expected to learn how to solve problems in a short time. Problem solution for an expert of your caliber extends beyond finding an answer. In this capacity, you are expected to offer solutions to problems from their root cause, in the process making sure that the problems are not experienced again.

- *Customer appreciation*

Wherever you are, customer satisfaction is always one of the key goals that companies demand of their employees. When customers are getting quality services, they are happy and look forward to doing more business with your company. Service delivery is at the top of all companies' to do lists. By getting your green belt certification, you are in a better position to attract more customers and enhance your appeal to their needs. When customers know that they can expect nothing but the best services from you, it follows that the company will enjoy amazing returns from business with the customers. You do not have to do too much to convince the customer to invest in whatever products or services you are offering.

- *Personal approach*

Green belt training is one level where you learn a lot about interacting with people around you. The training should boost your self-esteem, given that you are taught how to speak confidently, propose and champion new ideas, and use the same to solve problems arising. Having the necessary interpersonal skills is important, given that you can use this knowledge to create an engaging platform for your team.

Most teams struggle because the members do not have an environment where they can interact freely and exchange ideas. Successful leadership also depends on how well you can listen to your juniors, encouraging them to speak to you freely without worrying about reprimand. This is how to create a successful and effective environment for your team.

- *Competitive career path*

As a green belt, you stand a better chance of advancing in your career. The certification makes you more competitive, especially when making a sales pitch or in a marketing campaign. You might not know it at the beginning, but this certification gives you an edge over the competition.

Black Belt

As a black belt, your focus is on structuring, measuring and analyzing the problems that your organization is going through. Compared to the other six sigma roles, black belts have distinct functions in the organization. Their primary role is providing leadership in the project. The kind of training that you go through as a black belt focuses on the use of lean concepts and incorporating statistics into management decision making.

You also have a better understanding of the team dynamics; therefore, you are in a position to understand your team and assign them roles according to their capabilities. While a lot of people in leadership barely consider the need for variation analysis, black belt professionals do. This is a concept that you will use for quality decision making.

Benefits of Black Belt Training

You might have personal reasons you are enrolling for a black belt certification. This training will equip you with skills and experience to achieve your personal and organizational goals. Here are some benefits of enrolling for a black belt certification:

- *Rising to new challenges*

If you want to be a leader, you must be able to realize and take advantage of the challenges that arise. Most of the time, leadership is determined by how well you address conflicts and manage challenges that come your way. This training equips you with new challenges concerning knowledge and leadership experiences.

- *Advance new opportunities*

Black belt training is one of the highest levels of training in Lean Six Sigma. At this level, you are capable of going after new, better opportunities. Assuming one of your career plans was to get a job as a project manager, for example, with this training, you can achieve just that.

- *Better communication*

Project management requires the best communication skills especially between the team leader and the team members. In your capacity as a black belt professional, you will have learned how to do that. Effective communication through the life of a project is mandatory if you are to succeed as a team and the entire organization.

- *Technical understanding*

Black belts usually have the best understanding of the DMAIC methodology that is used in Lean Six Sigma. You get a deeper understanding of the process and how to perfect it in each instance where applicable in your place of work.

- *Embrace true leadership*

At this juncture, you have all it takes to become a true leader. One of the skills you will learn is how to establish a link between the capabilities of your team and the requirements of your organization. This will come in handy during staffing, so you can match the right people to the right tasks and roles within the organization.

- *Understand the competition*

We live in an enterprising world where businesses are constantly getting killed by the competition or market forces. A black belt professional should be able to understand the competitive forces in the environment within which the business operates. This is not just about identifying the competition, but also about winning over their market share while improving the overall value of your brand.

- *Customer relations*

Professionals in your capacity are often the link between the company and the customers. You can become a personal brand affiliated to your organization. Whenever things are not right with the company, everyone will look to you, especially the customers. They look forward to working with someone who will safeguard their interests in the company, particularly dividend-earning customers.

- *Management liaison*

Your role as a black belt professional is also to be the link between the rest of the team and the management team. In whichever organization you will work in, you will assume the role of a project manager. This naturally makes you a people person. In this regard, you will be relied upon to communicate effectively the needs of the management to the rest of the team and to do the same for the team to the management. You must not come off as someone who has a bias toward either of the factions.

Master Black Belt

A master black belt is a professional that is tasked with strategic Lean Six Sigma deployment in any organization. This is someone whose role will include improving all activities within the business, including organizing customers and suppliers.

A master black belt is the highest ranking Lean Six Sigma level, and it comes with certain responsibilities that cannot be accorded to lower level employees. As a master black belt, you are expected to offer support to all the Lean Six Sigma levels below you. Your mentorship should enable them to learn how to use the tools available at their disposal to solve problems in the organization.

You will also be required to develop and deploy organizational metrics. From time to time, the company can request you to look at the Lean Six Sigma curriculum as applied in the workplace and revise it to suit the growing needs of the organization. You, therefore, will be required to link up with different agencies that are responsible for external tutelage to make sure that the company stays on course with Lean Six Sigma training.

One of your strongest traits should be reaching out to a network of master black belts. Through networking, you can learn so much from them, which will help you learn how to put measures in place to improve the position of your organization in as far as competition is concerned.

Expectations of Master Black Belts

Given that this is a high-profile position in any organization, there are prerequisites that companies need for someone in your capacity. The following are things that you should have accomplished before you can be hired in such a position:

- At least five years of experience in business management, and no less than two years of black belt mastery.
- You should have in-depth knowledge of deploying the Lean Six Sigma tools in the organization, including making adjustments where necessary to suit the needs of the company.
- You must portray experience in deploying Lean Six Sigma projects, with tangible results given your desired position.
- You should be a leader. As a leader, you must have managed cross-functional projects, and show astute

leadership where your subordinates can follow your lead without having to be coerced into action by your authority.

- Change management is common in Lean Six Sigma deployment. With this in mind, you must also show your efficiency in managing change in organizations.
- Given that you will have lower level Lean Six Sigma professionals working with you, the company will need you to offer training exercises especially for green belt and black belt employees.
- You should be a strong communicator. This must be seen in your verbal and written communication with those who work around you.

Benefits of master black belt training

As a master black belt, some benefits should accrue to the company under your role in the organization. As a student of Lean Six Sigma, you can look forward to the following benefits from the training experience:

- Devotion to Lean Six Sigma

A master black belt is someone who has experienced all the levels of Lean Six Sigma, and you are at your peak. For this reason, you understand and breathe Lean Six Sigma in all you do. Therefore, you can lead, train, and ensure that your company has qualified mentors who can guide the rest of the staff members through their Lean Six Sigma journey.

- Oversight role

One of the tasks that are expected of you as a master black belt is to oversee the Lean Six Sigma process in the organization.

While you might need to reach out to external parties from time to time for Lean Six Sigma, you have the knowledge needed to oversee the internal Lean Six Sigma processes to fruition.

- Project leadership

Companies need leaders to champion and manage their projects if they are to succeed. In your role, you have the necessary skills to make sure all the projects are running according to plan. You are also required to interact with others and help them wherever they might be struggling in the implementation process. Your guidance should help a lot of employees in the course of meeting their requirements with maximum efficiency.

- Managing variation

As a master black belt, you have excelled at all the levels of Lean Six Sigma before this. You, therefore, understand the importance of Lean Six Sigma in your organization and how it should be implemented. Most of the junior management will refer to you when they need help in overseeing their projects. As an overall mentor, you are expected to support all the junior Lean Six Sigma levels and make sure that variation, defects, and errors that threaten to derail the business processes are eliminated.

- Multifaceted approach

A master black belt has the necessary skills to stand in as a teacher, as a consultant, and as a business leader. These are roles that you must perform across the board with ease. It is from your execution of these roles that other Lean Six Sigma employees can also use this opportunity to improve their current position by learning from you. In essence, you are a role model.

- Career opportunities

As a master black belt, your position and role is highly sought after in many industries. Once you have attained master black belt status, you can choose to stay with the company or step out and seek greener pastures. A lot of people these days walk out of employment and leverage their networking skills into consulting, and succeed while at it.

- Beat the competition

Master black belts have a penchant for success. These are experts who take their time to focus on ways of improving business processes while eliminating waste along the way. For this reason, you will succeed in putting the company in a profitable and competitive position in the market, and probably the industry as a whole.

Importance of Lean Six Sigma Certification

Lean Six Sigma methodologies can improve your life and career, leaving a lasting impact in your future. Undertaking this certification process means that you are committed to improving your analytical and business process skills, which is something most companies look for. A lot of companies are struggling to stay afloat, yet all they need is someone who can refine their business processes and improve performance for them. Bearing this in mind, mastery of Lean Six Sigma will set you a cut above everyone else. Starting a course on Lean Six Sigma is a brilliant idea. Here are some reasons that should give you a confidence boost as you start your classes and hope to get certified in the near future:

- Quality management experience

In quality management, there are many certification schedules that you can consider. However, Lean Six Sigma provides invaluable experience that you might not get elsewhere. In Lean Six Sigma, you deal with some of the most important factors that are at the core of business survival: managing and eliminating waste in business processes, customer satisfaction, improving efficiency and fostering teamwork.

A certified Lean Six Sigma professional is an asset to any organization, even before you join the workforce. This also explains why a lot of professionals who have delved into consulting have succeeded while at it.

- Improved earnings

One of the best things you will enjoy about Lean Six Sigma is the salary outlook. Lean Six Sigma professionals are some of the highest paid employees all over the world. With this certification, especially if you make it all the way to master black belt level, you should command earnings in the region of $100,000 and above.

- Leadership and managerial mentorship

As you go through the Lean Six Sigma process, you will learn to become a leader, a mentor and a professional confidant in the workplace. You become one person that someone can approach with a work-related problem, and expect you to address it professionally. As a mentor, there is so much that you bring to the organization.

You have the technical capacity to manage costs, improve the revenue outlook and ensure compliance and improvement in business process efficiency. You are not just learning how to become a Lean Six Sigma expert or how to implement the methodologies, but you also become an agent of change. Your leadership will shine through when management is discussing process improvement, service delivery to end users, and improving product quality. While companies seek value addition in their products and services, getting you on contract should be the first sign of value addition to the organization.

- Industry compliance

Lean Six Sigma is strict on product service quality delivery. You know how to reduce errors at work, and eliminate discrepancies within your processes. Bearing this in mind, you have the knowledge, skills and technical know-how to get your company compliant with industry requirements, especially with ISO certification and other international standards. Should the company improve their profit margins, a lot of this effort should be directly attributable to your input.

- Versatile experience

When learning about Lean Six Sigma, you will realize that you are not confined to one organization or industry. The Lean Six Sigma methodology was designed to be used in all industries. Your expertise can help companies improve and manage change accordingly. Whether you are in the marketing industry, financial services, aerospace, manufacturing sector, HR, IT, or electronics you are an invaluable member of whichever team you are introduced to.

One of the important things about your position is that as an agent of change, companies need you from time to time, especially when they need to restructure. Restructuring takes time, and some companies barely get through it unscathed. This is why your guidance would be highly appreciated. Besides, you should help the organization improve things like revenue earnings, cost management, waste elimination and increase motivation, helping the employees feel they are part of the process. These are important things that determine success within the organization, and a strong competitive advantage.

- Efficient business process management

Business process management requires a keen eye. A Lean Six Sigma expert has the know-how to analyze, measure, control and improve business processes. You are also aware of how to identify the characteristics of your organization and how you can influence these for future success. More often, you will need to review the entire policy and business procedures to identify flaws and make recommendations on how to deal with them.

Lean Six Sigma professionals are capable of improving the quality of mentorship processes and monitoring the processes to identify flaws and improve them. While working with other members of the management team, you are expected to ensure the company barely deviates from the mean, ensuring that they stay on course with all projects.

- Waste elimination

One of your strengths as a Lean Six Sigma expert is waste elimination. You become a critical part of the business owing to your ability to identify and eliminate errors within the business processes. You can identify repetitive errors.

Waste elimination goes a long way, in that it helps improve

customer satisfaction along the line. How does this happen? Customers are dismayed when the business processes result in substandard products or service delivery. Bearing this in mind, remember that your role will include identifying wastes and errors in the organization and eliminating them. Apart from that, you will follow through and institute measures that improve efficiency in all business operations.

In your capacity, some of the tasks that you will be expected to handle include reducing cost overruns, scheduling delays, unnecessary spending on overheads, improving the complaint resolution process, and minimize invoice errors or eliminate them.

CRITICISMS OF LEAN SIX SIGMA

A discussion with most professionals today will reveal one thing, most of them know about Lean Six Sigma. Enterprises today have leveraged and implemented the strategies learned to help them improve their productivity over the years. Given all the benefits that accrue to companies and individuals that have mastered Lean Six Sigma, it is surprising that there are factions that are skeptical about Lean Six Sigma.

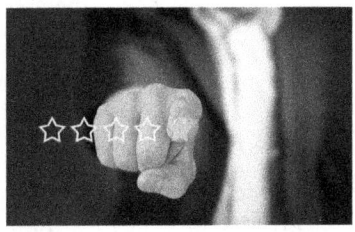

Other than those who have developed cold feet, there are others who have simply failed to harness the full extent of Lean Six Sigma and, as a result, ended up having nothing but a bad experience. Criticism is normal in any society. Some companies have given Lean Six Sigma a wide berth, even as others are scaling the heights of productivity through it. It is important, therefore, to learn some of the reasons that they give for their stand, and perhaps look at their experiences in a bid to learn more about their predicament.

A lot of entities and individuals who have spoken out against Lean Six Sigma in the past have based their sentiments on misconceptions. The reason for this is because while it is acceptable that it is difficult to please everyone, the majority who have tried Lean Six Sigma are enjoying the best outcome, and this is something worth appreciating.

Reasons Some Companies Do Not Appreciate Lean Six Sigma

Lean Six Sigma is about eliminating waste and in the process adding value to the service or production process. For this reason, therefore, it is almost absurd when an entity stands firm against a process that would speed up the process of getting things done, and saving them time. Essentially, these are companies that are standing firm against gaining a competitive advantage in the market. The following are some of the reasons that have been given by professionals and companies alike, for not preferring Lean Six Sigma.

Fad Practices

You must be aware of Total Quality Management (TQM). Many other similar processes have in the past been shunned as fads, which bear little or no effect on some organizations. Therefore, any company or professional who might have had a firsthand experience with these would have cold feet when it comes to Lean Six Sigma.

Sadly, this is a misconception. Lean Six Sigma is not a fad. Lean Six Sigma has stood the test of time. It has been used by many companies, and individuals and they have succeeded. The earliest mentions of Lean Six Sigma trace back to the nineteenth century. Lean Six Sigma was implemented by industrial luminaries like Taichi Ohno, Joseph Juran, Edwards Deming, and Henry Ford.

As time went by, Lean Six Sigma has adapted and been assimilated into modern business practices. Companies and professionals who embrace Lean Six Sigma do so because of the focus on ROI, persistent use of data and analytics in decision making, and the fact that it is a customer-centric approach to streamlining efficiency in the workplace. These are three tenets that every management department pursues.

Alien Concept

Believe it or not, some people are yet to hear about Lean Six Sigma. This reason is valid too. You cannot embrace something that you do not know about. According to most people, Lean Six Sigma is one of the most popular practices that will guarantee total quality management at work. However, when you consider the business sphere, it is not as popular as it should be.

Most people will learn about Lean Six Sigma through a Google search, and then from there, they begin their foray into the learning curve. If you are one such person, or your bosses share the same sentiments, perhaps you can guide them into learning the basics of Lean Six Sigma, and how they can take steps to implement it in the workplace. On your part, you can find webinars and lots of other useful resources online. These are useful tools which can be helpful as you try to learn more about Lean Six Sigma and figure out how it will influence the rest of your career.

Lack of Time

Lack of time might be a valid reason depending on how it is

presented to you, or it might be a lame excuse. Many are the instances where you will discuss Lean Six Sigma with someone, only for them to retort about how busy they are to spare time for it.

Time is one of the most important and most misused resources in any organization. Most people realize the value of time when they suffer the consequences of wasting time. The problem with wasting time is that it is a resource you can never salvage, unlike some materials. If you find yourself presenting this excuse as a reason not to consider Lean Six Sigma, it is time you re-evaluated your options and invested more time in learning how this process can help you change for the better.

Cost Considerations

Like time, one of the other reasons that people front for ignoring Lean Six Sigma is that they do not have the financial muscle to support the implementation of the program in their business. Unfortunately, most individuals barely recognize the fact that Lean Six Sigma is a process whose implementation barely requires major overheads.

For many years now, small entities have introduced Lean Six Sigma to their workplace in subtle steps. A day of yellow belt training can make a very big difference. In Lean Six Sigma, you learn the fundamentals first then build on that along the way. The solution here is to change your perception of Lean Six Sigma. Instead of looking at it as spending, you should see it as an investment. The knowledge and skills learned will be worth so much more in returns five or ten years down the line, especially when you look at project management and the general business environment.

Inferiority Complex

Many companies suffer an inferiority complex, thinking they

are too small. This is a problem with many small and medium-sized enterprises. In most cases, such companies believe that Lean Six Sigma is not for an entity of their size, but for giants in the industry, which is wrong.

A lot of these companies complain that their finances are tied up in receivables or inventory, even though you are certain the company is turning profits. Because of this notion, such companies can barely scale up and grow their businesses from the size they presently are.

Lean Six Sigma has been proven effective for small and large enterprises alike. A lot of enterprises have used the Lean Six Sigma practices to grow their businesses from strength to strength. Since this is a process that is focused on growth, streamlining processes customer satisfaction, it does not matter your present size. What matters is that when it counts, Lean Six Sigma can help you turn your small firm into a medium-sized entity, and before you know it, a large company.

Lean Six Sigma is important even for small companies because waste is always present, especially in a company that has not implemented it yet. The DMAIC strategy shows you how to identify wastes in your processes, and how to get rid of them.

We Are in the Service Industry

You will come across people who assume Lean Six Sigma is all about the manufacturing and production industry. This is a misconception that is easy to understand, but at the same time easy to dismiss. The reason behind it for a lot of people is the fact that history shows companies like Ford and Toyota were at the helm of developing Lean Six Sigma.

While it is true that Lean Six Sigma started as an important process in the manufacturing sector, it has since evolved and is adaptable in almost any industry that you come across. Companies like Bank of America, AT&T, and even hospitals have implemented Lean Six Sigma in their operations effectively, and they are reaping amazing benefits.

The principles of Lean Six Sigma apply to all sectors, even in the service industry. A careful look at the service industry reveals a lot of resource waste, more than the manufacturing industry. This is because, in the service industry, all the services are invisible. There is no tangible evidence to show service delivery. As long as you are in a position where you can collect data about the processes in your work environment, you have the prerequisites for introducing Lean Six Sigma processes into your operation.

Complex Concept

Lean Six Sigma uses scientific and mathematical principles. You will have to use advanced mathematics and statistics from time to time. The problem with this is that a lot of companies barely need or use these procedures in their work. Therefore, using Lean Six Sigma would be forcing them to get out of their comfort zone.

Unbeknown to most of these entities is that with Lean Six Sigma, the complexities are not always an issue. In most cases, you can identify waste by getting a second opinion of your processes. It could be from someone in a different department that seems to be performing better than all the others.

Instead of worrying about the challenge of introducing mathematical problems, you can also draw a process map on a whiteboard. On the process map, you can discuss with your team and identify bottlenecks and redundancies in the process, which create waste. More often, all you need to do is challenge your team to go the extra mile. You must keep them motivated, however, for this to work in your favor. It can be as easy as communicating clearly what your customers need. This is one way of encouraging them to share new ideas on how to go about the task at hand. You would be surprised at the information that you receive from them.

We Are Comfortable with Lean

While discussing Lean Six Sigma, you will realize that there are entities that have practiced Lean separately from Six Sigma. This works for them. However, this should not mean that there is no room for improvement. In such cases, the plan might be implementing Lean effectively, and then after that, they can bring in Six Sigma.

This is a good plan, but that is just it. You need something better. While this might work for you, you will spend more on resources to implement the two principles in isolation. By design, Lean and Six Sigma are supposed to work together. They complement one another. While Lean is about improving efficiency in throughput and speed in the business, Sigma is about improving product and service quality by eliminating or reducing variation and defects.

Instead of implementing either of the principles in isolation, you should combine them to give your business the best of both worlds. Focusing on Lean alone will see you compromise on quality. This would be a waste of your effort and resources over the long-term. The same applies when you focus on Six Sigma and leave out Lean.

We Have Done This Before

Mistakes happen. Even the best strategies can be implemented and fail. Does this mean that you should quit? Remember that quality management strategies like Lean Six Sigma have been in place for years and have delivered tangible, realistic results. Therefore, if you have tried it in your company before but failed, you should reconsider your options, evaluate all possibilities, and understand the reason you failed the first time.

More often you will realize that you had a good plan, but a few elements in your organization hindered it from succeeding. Perhaps you did not have the right personnel to carry out the exercise. Maybe you did not have the necessary equipment, or your timelines were not realistic. Once you have identified the reasons the earlier plan failed, as yourself what you have done since then to make things different.

Going back to such a point in time requires a clear conscience, so you can reflect on the reasons for failing, and make sure you understand why the process failed. Be honest if you are to learn from your earlier mistakes.

Whatever the reasons you failed when implementing Lean Six Sigma earlier on, you owe it to yourself to give it another try. Your business, employees, and customers deserve better. Address your current problems, and if they go way back, study the chronology of the problems and see how you can use the knowledge of Lean Six Sigma to make things better. Remember that to implement Lean Six Sigma and succeed, and the plan will only be as good as the people you recruit to spearhead its adoption and management. Therefore, ensure you get the selection right.

What If We Fail?

Fear of failure or the unknown holds a lot of people and companies back. When you are afraid, you get stuck and cozy up in your comfort zone. The problem with comfort zones is that you never realize when the game is changing around you. The competition could be getting bolder, but you remain oblivious to the fact.

One of the reasons companies are afraid of failure is pride. No one wants to be seen failing. In retrospect, you end up paralyzing your business because you are afraid to venture into untested waters. If you keep waiting for other companies to test the waters before you make a move, you will end up scrambling for what is left of their customers after they have lured the crème de la crème away from the competition.

For your organization to thrive, you must be bold in your business approach. Entrepreneurs are risk-takers. Take a chance and implement Lean Six Sigma in your workplace. Talk to experts; discuss these prospects with your close circle of professionals. Introduce the idea to top management, and challenge them to give it a trial run

CONCLUSION

Contrary to what most people believe, Lean Six Sigma is not just applicable in the manufacturing industry. It is a process that can be used effectively in most organizations, even in the service industry. A lot of companies have embraced Lean Six Sigma today and are doing well against their competition.

There is so much that you can learn about Lean Six Sigma. It is always wise to take a wider perspective if you are open to learning. Many people have criticized Lean Six Sigma in the past, but most of these critics often look from the outside-in. They barely have an idea of what goes on in Lean Six Sigma, because most of them work with assumptions and myths. Lean Six Sigma is a factual process. As such, any criticism should be leveled backed with facts. We have outlined some of the common criticisms that you might have come across against Lean Six Sigma, and debunked them with factual information on quality management. When you venture into Lean Six Sigma, therefore, you should do so with a clear conscience, and purpose.

Implementing Lean Six Sigma is not one of the easiest things, especially for someone who does not have sufficient resources to work with. More often, you will be prone to making mistakes. Even those who have sufficient resources at their disposal make some mistakes during the implementation stage. Common implementation mistakes have also been discussed, and how you can challenge yourself to overcome them and succeed with Lean Six Sigma.

It is easy to look at Lean Six Sigma from the perspective of what the company can gain from your expertise, and overlook the personal contribution that these classes will have in your life. On your career growth path, Lean Six Sigma is something that will make you a desirable employee in your organization, or take a leap of faith and consider a career in consulting. Whichever the case, your certification will get you places, more so if you have attained master black belt status.

The best way to succeed with any project is to ensure you have all the interested stakeholders backing the project. Project backing, especially by management is mandatory if you are to get their confidence in the project, and motivation to pursue the project goals. In case you are worried about how to convince your management that Lean Six Sigma is the way to go, read on and you will learn a few tips to get them to have a change of heart especially if you are facing hostility from management, but for the good of the company, you have to employ Lean Six Sigma, and you will also find herein some useful information on how to win them over and prove to them that supporting your Lean Six Sigma cause is the best thing for the entire team.

The information contained in this book should act as a guideline that will set you on the right path to success. Even if you get to master black belt level, nothing comes easy. You will still encounter a few challenges, and it helps to have a backup plan, which can be useful when you need to reboot the plan and start from the basics. This book gives you the foundation of Lean Six Sigma, upon which you can chart the course for your future in management.

REFERENCES

Aalst, v. d. 2016. *Process Mining: Data Science In Action.* Columbus: Springer.

Allen, T. 2018. *Introduction to Engineering, Statistics and Lean Six Sigma.* Columbus: Springer.

Bpir.com. 2019. *Total Quality Management History of TQM and Business Excellence BPIR.com.* [online] Available at: https://www.bpir.com/total-quality-management-history-of-tqm-and-business-excellence-bpir.com.html [Accessed 30 Jan. 2019].

Coleman, B. (2014). *The Customer-Driven Organization: Employing the Kano Model.* London: CRC Press.

George, L. M. 2002. *Lean Six Sigma: Combining Six Sigma Quality with Lean Production Speed.* Chicago: McGraw Hill.

Harvard Business Review. (2019). *How to Deal with Resistance to Change.* [online] Available at: https://hbr.org/1969/01/how-to-deal-with-resistance-to-change [Accessed 3 Feb. 2019].

Isixsigma.com. 2019. *Reducing WIP at a Frozen Food Manufac-*

turer | iSixSigma. [online] Available at: https://www.isixsigma.com/methodology/dmaic-methodology/reducing-wip-at-a-frozen-food-manufacturer-part-1-of-2/ [Accessed 3 Feb. 2019].

Martin, J. W. 2014. *Lean Six Sigma for Supply Chain Management*. McGraw Hill Education.

Process Excellence Network. 2019. *Driving an Effective Customer Experience Using Lean Six Sigma: An Interview with Doug Burgess of Xerox*. [online] Available at: https://www.processexcellencenetwork.com/lean-six-sigma-business-performance/interviews/driving-an-effective-customer-experience-using-lea [Accessed 3 Feb. 2019].

Shankar, R. 2009. *Process Improvement Using Six Sigma: A DMAIC Guide*. Wisconsin: ASQ Quality Press.

Sixsigmadaily.com. 2019. [online] Available at: https://www.sixsigmadaily.com/case-study-tire-manufacturer-dmaic/ [Accessed 3 Feb. 2019].

Trends, S., Relationships, 7. and Alton, L. (2019). *7 Strategies for Better Managing Client Relationships—Small Business Trends*. [online] Small Business Trends. Available at: https://smallbiztrends.com/2017/07/client-relationship-management.html [Accessed 3 Feb. 2019].

LEAN SIX SIGMA

THE ULTIMATE INTERMEDIATE GUIDE TO LEARN LEAN SIX SIGMA STEP BY STEP

INTRODUCTION

"In my wildest dreams, I would not have imagined how much money you could save," said Todd Graham. "Saving money for the company was really fun for me."

Todd has been with Johns Manville (JM), for 18 years. Except for a short time as a car mechanic, JM is the only company he has worked with since his apprenticeship. His career as a production / machine operator ran smoothly into the late 1990s until JM introduced so-called, "teams to reduce variations."

Although he had only a secondary school diploma and some advanced courses, Todd has always been interested in process optimization. Therefore, he applied as a team member and was accepted. The team members were released from their previous duties, received training in statistics, and started their work.

"Every 'Variations Reduction Team' made some progress," Todd said, "but in actual fact, the teams were just cluttering factory workers who somehow got through each other. Nobody really understood the big picture."

That would change soon. In October 2001, Todd accidentally read a JM newsletter in which the Company announced the launch of something with the name Six Sigma. Todd recalls that only a few details were mentioned in the newsletter, but it was said that the company was looking for volunteers who would become so-called, "Black Belts" - people who were supposed to manage projects at company locations. The article further stated that the Black Belts would take at least two years to complete the job.

That sounded very interesting to Todd, but he was not sure at first, if he really wanted to get involved in the Six Sigma initiative. However, one month after reading the article, he was directly approached by management and asked if he would be involved in the program. He agreed.

"What ultimately persuaded me to become a 'black belt' was the massive support that came from corporate headquarters," Todd said. "I was convinced that we would now receive much more support than we had previously had in the 'Variations Reduction Team.'"

As it turned out, he was right. "The top-down support worked really well," he said.

Soon after Todd was officially released from his regular work assignments, he began the first of five weeks of training.

"In the first week, we learned everything about leadership," he explained. "Then we had another four weeks of additional training at intervals of one month each. That means we had one week of training, then we went back into the company and worked on a project. Then we met for the next training week, went back to the project, and so on. "

Todd described his work as part of the training project as an experience that had opened his eyes. "We could not imagine that it would be so difficult to find usable process data," he says.

In the process in which he was involved, various raw materials were melted and processed into glass. The goal of the team was to make the process and the product more consistent. Therefore, the team would need to determine how well the materials were mixed before and after transport.

"The analysis of the process examples could take two to three weeks," said Todd. "As a result, the data I looked at each day reflected the status of three weeks ago. Of course, that did not help me much to manage what was happening. "

Ultimately, his team was only able to solve a part of the existing data problems and achieve only about 40% of the original project goal. "I learned from this that projects have to be carefully delineated.".

Finally, to help with this, JM's insulation division brought together all of its black belts in a two-day project selection and review seminar.

After completing his training in April 2002, Todd's working life became more interesting and a real challenge. "The benefit is that using Lean Six Sigma is a lot of fun, especially if you enjoy working with people and statistics and understand process operations." But in the beginning, you have to learn a tremendous amount. I understood most of the statistical tools and concepts very quickly, even if a lot of content was presented in a very short time. "

His first year meant an additional challenge for him, he reported, because high demands were placed on the JM Black Belts. "We did most of the project work ourselves and also tried to coach and advise the other people involved in the Six Sigma initiative. Today, the company increasingly trains Green Belts. These are employees who receive basic training - to a lesser extent than the Black Belts - and then participate in projects."

This had the positive effect of allowing the Black Belts to focus more on their work as internal consultants, guiding the teams, and helping them to use the right data tools.

"What was really great was the company-wide good results. I think the net profit from last year's Lean Six Sigma projects is somewhere between $ 3 million and $ 4 million, "said Todd. "If you know where to look, you can find a lot of easy-to-harvest fruits."

Todd's story is typical of people who have worked with Lean Six Sigma as Black Belts. For them, the experience was a challenge and, in many ways, rewarding.

Like Todd, some readers of this book are likely to take part in extensive Black Belt training for which they give up their "normal" job. Others may be expected to complete a basic training course on Lean Six Sigma. With all these offers, you have to decide for yourself how actively you want to support Lean Six Sigma in your business.

The purpose of this book is to give you enough background information to enable you to make that decision for yourself.

Arguments for Lean Six Sigma

Our best argument to convince you, not just to read this book, but to seriously engage with the topic, is that Lean Six Sigma almost only brings benefits. We know that many companies have made other efforts in the past to bring about (process) improvements and have failed.

Therefore, it is not surprising that many people are skeptical about Lean Six Sigma.

But even in the worst case, if all efforts have been in vain, at least improve your job qualification through the structured nature of the Lean Six Sigma training and education modules.

The second-best argument for joining Lean Six Sigma is that this collaboration brings you tremendous benefits. By using Lean Six Sigma in your own work environment, you can effectively help your business to become more profitable. Every company relies on some sort of customer base in order to bring in profits, making customer satisfaction paramount. Through the implementation of Lean Six Sigma, you are capable of analyzing the key attributes of customer satisfaction in order to identify those factors that affect customer fulfillment. This will help the company retain their customer base.

Beyond customer satisfaction, companies need to look closely at their business processes. Through SMART goals, and the implementation of Six, employees will be able to better manage their time and their projects. The data principles that Six Sigma creates will also help these employees break through the points in project successions that often create delays. Time is money, no matter what way you look at it and efficient processes can help your company reach those project financial goals.

One of the most frustrating parts of the management of a project, are the changes in project scope. Most of the time, these changes come out of nowhere, when a problem is identified during the actual progression of the project. Six Sigma allows the project manager to create watch teams that are put in place to find these issues before the team reaches that point in the project. This allows the team to focus on problem solving instead of problem identification. These solutions, ahead of time, should help to increase the probability of meeting project deadlines.

Even outside of specific, customer based, projects, the company strategy as a whole can always be developed. Implementation of Lean Six Sigma will allow a company to improve upon their inner workings based on their company's business objectives. Six Sigma will be able to improve these processes, saving money, helping to better retain employees, and create a more seamless production from top to bottom.

Lean Six Sigma doesn't just improve the company as a whole, but the productivity, skill set, and motivation level of the employees within its walls. Through implementation of the program, employees will have access to valuable resources that will improve upon their decision making abilities, problem solving skills, and team skills. With these improvements, oftentimes, you can cut down on the amount of people within the project chain. These eliminations will not only improve quality of life for the employees, but will eliminate all kinds of waste, save time, and make the produced outcome more desirable.

In Your Own Words

We talked to several people employed by companies where Six Sigma or Lean Six Sigma was just introduced. Here are some remarks from these people about their experience of implementing Six Sigma and Lean Six Sigma:

- Heather Presley, Fort Wayne Town Hall:

"Do not mess with us unless you're totally convinced that you're going to get the bottom line. There were days when I saw the process and thought that it would never be possible to achieve the results of the kind we had planned. It really required tenacity. But you do not have to fight it alone. If your business is always focused on big profits, there are likely to be a few failures. But if you train

employees in terms of process improvement think, they will become better employees."

- Ashish Merchant went straight from college to Western Union a few years ago. He works there in the area of international money transfer, a fast-growing business unit of Western Union. In 2002, he had the opportunity to participate in a Green Belt training course.

"I quickly realized that this was not the 'Einstein theory.' Six Sigma contains a lot of general logic and is also a structured approach. Some people teach Six Sigma as if it were the Evangelism. But that's not a good approach. There is no need to always fight through the whole. Just use the areas that are good for your project and take it further. Use your common sense."

He also found that he can apply what he has learned in his daily job. Productivity in the workplace can be significantly improved, even if only simple methods are used.

- Barry Shook, manager at Xerox, moved from production to the service area of the company. He has learned that the implementation of Lean and Six Sigma techniques has not only improved the way he works in this area, but has also created some competition in his work environment.

"One of our customers recently decided to introduce Six Sigma into their company. That's why he was looking for suppliers who know what Six Sigma means and have integrated Six Sigma into their own environment. In order for each and every one of us to be successful, it is our responsibility to provide only first-class services to our customers. The key is not to react, but to act. Lean Six Sigma provides you with the steps, tools, and methods to reach this level."

We cannot guarantee that your company's efforts in the area of Lean Six Sigma will be successful or lasting. But with the profits that both companies and their customers see, it will be very difficult to get out of this area.

In municipalities, for example, it is likely that the entire management team changes every four years (by choice).

But as Heather Presley notes, "Even if the city decides not to continue practicing Six Sigma, it will remain. Why? Because the people to whom we offer our Services, in general, are already at a much better level of service on the part of the city. If it worsens again, they will notice. "

Regardless of the extent to which you will later address Lean Six Sigma, as you read this book, you will gain insight into methods and concepts that have helped many individuals make proven positive changes to their jobs.

The Benchmarking

Benchmarking is a very specific approach to compare its company, its organization, and its processes with its partners, and, in the best case, with its competitors. Benchmarking will enable the company to find those process gaps that aren't likely seen on a day by day basis. The goal from there would be to identify good practices that can be deployed internally in order to grasp ahold of the competitive advantage within the companies sector of business. But the approach can only work if you first define an adequate metric.

Benchmarking In Practice

Benchmarking is a specific tool that is often used on an ongoing basis within an organization. Oftentimes it is practiced by someone specifically certified and experienced in the process, but can also be broken down by company sector and used by the professionals within the different departments. There will be performance metrics provided, including fifteen to twenty core values that the initiator or department will use to grade the process against. It is used as a guide, a reference, and ultimately a report card.

Benchmarking can also be used for specific employee performance. They will include the same process but will also have a specific set of time sensitive goals to be completed. Within the timetables the employees will receive specified milestones that assists them in proceeding through the process in a manageable manner. After all, you are not testing them to find things wrong, you are simply attempting to find where you can help them improve. By setting them up for failure, you are ultimately creating a hole in the actual process itself.

Benchmarking Applications

Benchmarking is a process with multiple uses and different applications. The results of a benchmarking exercise will also be a valuable lesson in a strategic analysis in addition to a SWOT study.

However, the quest for good practices does not kill all the organizational innovation reflections. Inspiring models to find one's own path will be far more profitable in a competitive world than trying to copy them to the letter. The role of the follower is difficult to sustain.

Benchmarking can, of course, be used to evaluate not only the design, manufacturing, marketing / distribution processes, but also the products made, the services provided, and the support and support activities.

The 7 steps of The Benchmarking Process

1. **What should be compared?**

It is a question of focusing on the areas of progress, the aspects that have been selected as areas for improvement.

1. **What is the target?**

We are looking for leaders who excel in the aspects selected in the previous step.

1. **What measures?**

Criteria and measures are specified.

1. **Collection of information**

No source of information should be neglected. The point of view of the customers is not the least of the information to be collected.

1. **Analysis**

Two questions: why are we better and why? Where are they better and why?

1. **An active approach**

Benchmarking is only the first stone of change; then we must take action.

1. **A new benchmark**

We are re-launching a new benchmark in the process of continuous progress.

Brainstorming

Definition: Brainstorming is a participatory problem-solving method based on the spontaneous creativity of the participants. In fact, it is the spontaneity that is sought. Let's see the approach, or rather the protocol, to follow if we want to achieve a result.

To begin, you would want to decide which technique to use during your brainstorming session. Oftentimes a specific method is adopted company wide and used over and over again. Some of these analytic brainstorming techniques include Mind Mapping, a visual process used to push the process, Reverse Order, finding the problem first instead of the solution, and Gap Filling. Gap Filling creates the scenario starting at the specific point in the project where the team currently sits, and then works its way out in both directions. Other techniques include the Drivers Analysis, SWOT, The List of Whys, and a plethora of other choices.

Once you have chosen the method, you want to fully engage the staff, and educate them on the process you will undertake. This can include best tips, step by steps, and templates that go along with the process. You want to make sure that your brainstorming session follows a structure to avoid getting off topic or going beyond the bounds of the project. You want to also appoint either one specific note taker or instruct everyone involved to do so as the discussion moves along.

These brainstorming sessions can last one meeting or months. It all depends on the problems or project being discussed, and the timeline inside of the overall project. Once these initial decisions have been made, you will begin the brainstorming session.

Running a Brainstorming Session

"A problem is a gap between the current situation and the

desired situation. This gap is strewn with known and unknown obstacles. His crossing is not easy." - Piloter.org

The protocol of brainstorming in practice can be either very structured or up for interpretation based on the needs of the company and the problem at hand. The team and any other vital members of the process gather together to discuss and implement their expertise in order to create viable solutions to the problem.

It is important to understand that not everyone feels comfortable blurting out their ideas. You want to make the meeting comfortable and you want to set rules. During these sessions, any and all ideas that are thrown out are noted without initial judgments being made. Participants within the session should be asked not to criticize or give negative remarks on someone's ideas. The employees need to feel free to bounce ideas back and forth and allow the process to progress naturally until a solution or solutions are collected.

Procedure

A brainstorming session takes place in groups, often split into tens, with a facilitator who directs the discussion and keeps the team within the parameters of the project. These teams can be split by department, expertise, or preferably, mixed together to allow different knowledge to be combined and converted into solutions..

The first phase, lasting as long as necessary but usually no longer than a couple of hours at a time, aims to collect a maximum of ideas, as described above. You want to have as many choices as possible in order to find the perfect one. When all sessions within the first phase are complete, those ideas are then sorted, reconciled, thematically organized, and put into a hierarchical form.

Often times an Ishikawa diagram, also called a Fish Diagram

as seen below, or an Infinity diagram is used in order to efficiently log the data and compare easily. During this entire process, the facilitator should only be acting in the form of a guide. The facilitator can be a newer team member learning the ropes, a manager, or even someone outside of the project.

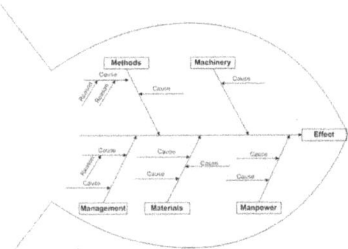

The 7 Steps to Successful Brainstorming

1. **Explain the Process**

It is important that each participant is fully aware of the rules and steps in the process. If one or more group members have never attended a session of this type, it may be helpful to do some hands-on exercises to fully understand the principles.

1. **Specify the Rules**

Every brainstorming session requires that you have a specific set of general rules that are to be adhered to, and monitored by the facilitator. Some examples of rules that can be used are not allowing criticizing or negative remarks, letting your imagination free, build on each other's expertise and thoughts.

1. **State the Problem**

Make sure that all participants have a correct vision of the problem to be solved. Before starting the session, leave a few minutes of reflection so that everyone can properly integrate the problem. There is no need to let spontaneity fly away. In fact, we privilege spontaneity over reflection. However, the moderator reserves the right to rephrase a proposal that seems ill-formulated.

1. **Write Everything Down**

The whiteboard, visible to all, is still the best tool. It works even in case of power failure.

1. **Cleanup the Ideas and Reformulate**

Eliminate or combine similar ideas formulated differently.

1. **Sort, Group, and Eliminate**

Bring the ideas together, and eliminate those that are impossible or irrelevant, as well as duplicates. The Affinity diagram, seen below, is the preferred tool for this "counting" operation.

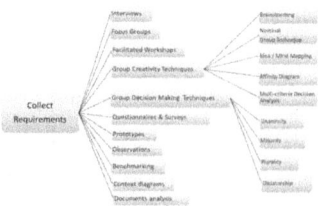

1. **Narrow Down to the Best**

Select the best ideas from the options provided.

Tip: Rather than turning the room into a fish auction, where only extroverts speak out, attempt brainstorming in writing. The ideas are then noted on a post-it, and then displayed on the whiteboard. On the other hand, brainstorming is an exercise in associations. It is also beneficial to use the key instrument, Mind Mapping to build ideas and solutions

Occasionally we meet participants who can be qualified "deviant" for this exercise.

3 Examples of Deviant Participants

- Someone who is given direction in advance, and only participates if someone's ideas go in that direction.

Someone who demands that all their ideas be preserved. They want to keep the property. They are obviously, at least to the person, much better than any other suggestions from their colleagues.

- Someone who appreciates only the peaceful atmosphere, and seeks to play the role of moderator when the exchanges become lively.

These types of participants should be coached in order to push them into a more open and free thinking Brainstorming session.

How to Succeed?

First and foremost, all participants within the session must have a connection to the issue at hand. Having someone outside, unless bringing a very specific education value to the session, will cause the session to lag. Within the session there need to be specific actions that are listed. These actions will act as guides for the participants, and allow for them to move forward as quickly as possible. Oftentimes, even with expertise, not all participants will have a full scope of the problem being discussed. It is vital to ensure these people are well educated on the subject beforehand.

What has been discussed to this point in the book gives a framework for the Lean Six Sigma program and many of the different places inside of the organization that it can be implemented. However, the point of implementing such a program is to ultimately increase your profit performance. From customer satisfaction to employee time management, it all comes down to the profit.

1

LEAN SIX SIGMA: CREATING BREAKTHROUGH PROFIT PERFORMANCE

The Six Sigma method, oriented quality, aims to reduce the variability of a process to tend towards zero defect. The Six Sigma method is based on an approach figured both on the customer's voice (surveys, etc.), and on measurable data (indicators, etc.), and reliable data.

This method is used in processes to reduce variability in production (or other) processes, and at the product level, and thus aims to improve the overall quality of the product and services.

The Six Sigma method encourages the company to adopt measurable and effective actions, to better satisfy its customers, to involve the teams, and often to improve its image.

What Is the Origin of the Six Sigma Method?

In 1986, the Six Sigma method came into being at Motorola, an American firm headquartered in the suburbs of Chicago. The Six Sigma® trademark has been registered by the firm.

The primary objective of the method was to increase customer satisfaction by improving the quality of product production processes, and thus the quality of products. The first field of application of the method was, therefore, industrial processes, but the method was then extended to other areas such as logistical, administrative, commercial and even energy-saving processes.

Why the Name Six Sigma?

The Greek letter sigma σ is the standard deviation; remember that the standard deviation is the square root of the variance in the mathematical sense. So "Six Sigma" is 6 times the standard deviation. The standard deviation can be likened to the dispersion of a process

In practice, the method aims to ensure that all the products that come from a process are included in an interval that is, at most, 6 sigma's away from the average products resulting from this process. Reducing the variability of the products in the process reduces the risk of seeing the product and / or service rejected by the recipient because it is outside its expectations or specifications. So, we try to improve the process until only the products that correspond to the expectations, to the specifications are delivered: to produce in an expected way from the first time avoiding thus the corrections, the retouches, the repairs and especially the associated costs.

Six Sigma and Total Quality

Every project implementing the Six Sigma method in an organization responds to a specific sequence and quantified objectives, for example, reducing production time by v%, reducing pollution by w%, reducing x% costs, increase customer satisfaction by y%, and increase profits by z%. Here we find the principles of total quality management (TQM), which is a quality management approach that aims to involve the entire company to achieve perfect quality by reducing waste and improving permanently the output elements. This approach is described in particular in Tom Peters' book "In Search of Excellence" which a French version exists under the name "the price of excellence."

An industrial process or service involves several repetitive tasks. This is the case, for example, for the production of a mass-produced piece. A part or service is what is expected if it meets several criteria, but the parts or services cannot be totally and strictly identical. The Six Sigma method aims to improve the process so that these products are all good; it is not a question of controlling the products, but of making sure that the process is reliable.

The Six Sigma method can be implemented in any kind of process and not just production; it is enough that the process performance is measurable. In practice, the limit of 6σ is difficult to reach, but some companies can search for 4σ.

Six Sigma Listens to The Voice of The Customer

For the method to be effective, it is first of all necessary to take into account the voice of the customer that is to collect and analyze the customers' opinions. So, setting up a Six Sigma approach involves, first of all, probing the customers on their real needs.

Once the customer's needs are clearly determined, the Six Sigma method will be able to question the ability of the organization's processes to deliver the expected product and / or service.

Improving Processes

In summary, the Six Sigma method is based, as you will have understood, on the concepts of customer, process, and measurement. It is therefore a question of measuring customer expectations, measuring the performance of the company's business processes against these expectations, using statistical tools to analyze the causes that affect performance, and setting solutions to correct the causes of non-performance, to use measurement tools to check that the solutions put in place have the expected effect on improving performance.

Thus, the Six Sigma method uses the DMAIC tool (Define, Measure, Analyze, Innovate, Control).Each DMAIC step has different tools. Here are some examples:

- **Define**: customer voice, surveys, SIPOC (Supplier Input Process Output Customer)
- **Measure**: analysis of measurement systems (Gage R & R, linearity, capacities, Ishikawa diagram)
- **Analyze**: detailed process mapping (for example, value-added analysis, hypothesis tests, ANOVA, χ^2, variance tests, experimental plans)
- **Improve**: Experimental plans, analysis of failure modes, their effects, and their criticality (FMEA, key)
- **Control**: experimental design, statistical process control, or MSP

Six Sigma Certifications

It is possible to pass certifications and learn how to properly implement the Six Sigma method. The most recognized are the certification of the American Society for Quality (ASQ), and that of IASSC (International Accreditation for Six Sigma Certification).

Climb the ladder of the Six Sigma pyramid. As in judo, the levels of expertise are represented by belts of different colors. Be successively White Belt, Yellow Belt, Green Belt, Black Belt, and Master Black Belt.

- **The White Belt** ("white belt"): First step to understanding Lean and the Six Sigma method.
- **The Yellow Belt** ("yellow belt"): Sensitized to Six Sigma issues, he contributes to the realization of an improvement project under the guidance of the higher levels.
- **The Green Belt**: expected to devote some of its time (often around 25%) to the conduct of improvement projects.
- **Black Belt** ("black belt"): team leader dedicated to full-time improvement (conducting projects, training Green Belts or other Black Belts) and must master the method as a whole.
- **The Master Black Belt**: mentor and trainer of Black Belts, a guarantor of the respect of the step, he is authorized to frame the Black Belts.
- **The Deployment Leader or Champion** (in France, "Director of Deployment," or more often "Director of the System of Excellence"): responsible for developing the strategy, the content of the training, the budgets, etc.

How to Combine Lean Approach and Six Sigma Methods

Lean is a method of systematic process improvement. It focuses its efforts on what is call the axes of intervention. This process includes taking a systematic approach to figuring out way to reduce cycle times, improve JIT, and improve not only fluidity but flexibility as well. These processes are going to help the company function better as a whole as well as from project to project. The teams will gain productivity and the process issues can be greatly reduced.

Six Sigma aims at the drastic reduction of all forms of variation below a range corresponding to customer satisfaction. This means, when the process is put to the test and receives a score card number, it will be decided what needs to happen in order to improve upon that score. The Six Sigma method is there to reduce all variation when it comes to customer satisfaction. There should be a steady score, that only waivers with your customers on special circumstances.

The juxtaposition of the two approaches, which are both process-oriented, makes it possible to manage the improvement approach globally by taking into account all customer expectations regarding quality, deadlines, and costs. When you put these processes together, and give the client a full and well spelled out project, as well as leading your project team with expectations, the customer score will be considerably higher than before.

Quality and Time, a Controlled Process

Quality and time are closely linked. Touch ups, returns, and rejects are major causes of the slowdown. On the other hand, the improvement of the delays within a process necessarily implies the systematic reduction of defects and manufacturing errors. In fact, all phases of the process that do not add value in the customer's sense deserve to be evaluated, deleted, or transformed for the majority of cases.

8 Types of Waste

Waste is something that all companies should look to reduce or completely erase from their businesses. Waste equates to less money, and unhappy clients. Just as wastefulness as a child is preached upon, wastefulness in a corporate setting can be extremely damaging to the bottom line as well as future client interactions. These are some of the most common wastes found within a business.

1) Defects and rejects as well as customer returns

2) Over-production, we do not produce on stock, i.e., without order

3) Over-storage, i.e., ban all stocks that are not essential

4) Waiting times, and all problems of synchronization between two activities

5) Useless movements as well as all the ergonomic errors of design of the workstation

6) The useless transports or the useless displacements of products, tools, or people

7) The useless treatments, which bring nothing to the product according to the expectations of the customer and increase the costs unnecessarily

8) The under-utilization of qualifications

How to Lead the Project Lean Six Sigma

Prepare the Project

Associating two radical methods like Lean Management and Six Sigma will not be considered without significant preparation. As with all change within a company, you want to be prepared on a financial and an employee level. It is often good to bring change management teams in during this shift.

The means anticipation of the budget, the availability, the skills, and the motivation of the key players, as well as the deadlines, will be up to the challenges. Each and every project should be looked into with significant detail, and the exact numbers should be processed from there.

The Eternal Question of Big Money

If it is not a case of emergency funding, it is better to avoid revising the budgets until the fall, or shorten the timings once the project is launched. A sense of patience when dealing with money is always pertinent since most financial changes take time to see the final results of.

The Scope of The Project

To offer the best guarantees to complete the project, it will be necessary to precisely define the scope of the project. The more well defined, better targeted it is, the easier it will be to define the feasibility, anticipate the ROI, and set the necessary budgets and resources.

Risk analysis and an impact study are also useful to evaluate the contributions in terms of collective learning. Risk Analysis are also used on occasion to complete a project or study in the preliminary stages. This will allow your team to understand their project based risk management as they go.

On the other hand, to create the dynamic that is essential to the pursuit of a wider ambition, it is preferable to place in the right place on the list of criteria for the selection of potential projects those whose results are significant for the whole. actors of the company and its partners.

Piloting the Project

The question of piloting, inseparable from those of the design of the dashboard and the definition of performance indicators (KPIs), is the central theme of this portal.

Collective Learning

The importance of the constitution of a project memory is

far too seldom appreciated. Neglected during the development of the Business Plan and the financing plan, the development of active and alive documentation is nevertheless the pillar of collective learning.

Do not lose the teaching of the first projects; it is better to prepare the following, increase the overall performance ratio in some way.

To constitute this collective memory, the interested reader will be able to refer to the topics of the project management: The Wiki and the elaboration of the documentation of the project, storytelling, and the project memory.

Change Management

For many years now, we have been talking about and writing a lot about the support of change, a particularly difficult question often associated with that of motivation.

Of course, when the term "change" is, in fact, a euphemistic way of trying to hide the terms "offshoring," "dismissal," or "downgrading with lost wages," there is little chance of succeeding without a crash.

Change and Cooperation

On the other hand, treated in a spirit of wide cooperation integrating by definition the communication, the participation and the responsibility, the exercise, always difficult, is much more in phase with success. We will have many opportunities to come back to this essential theme.

The post-project is the second difficulty of continuous improvement initiatives. How can we guarantee that the company is well placed on a new launching pad and has completely broken with past habits? This is where the question of change is revealed in all its dimensions. If the enthusiasm and successful crossing of successive difficulties during the project are stimulants, real drivers of change, the desire to regain a position of comfort associated with past habits once the trip is completed, will be the main threat.

It is then necessary to make every effort to avoid going back (with, among other things, the implementation and deployment of a measurement and control system based on distributed dashboards). In fact, change must be considered as much in its cultural, organizational or technological dimension.

It concretely develops the implementation of 6 Sigma. It concludes with an excellent chapter: "the roadmap for change" describing in 5 steps the key points of the "final change" approach.

How to Run the Lean Six Sigma Project for Services?

Lean Six Sigma and The Activities of The Tertiary Sector

The processes of tertiary organizations, service sector, are, in any case broadly, little different from the processes of the production industries.

Just as dependent on the voice of the client, the organizations of the tertiary sector, be it the services of type financial institution, the banks or the insurance companies, the administrative departments of the industrial enterprises and the public administrations, are subjected to the same imperatives of quality, delay, and profitability.

New Projects

The Lean Six Sigma approach applied to service activities, without being fully traced back to industry practices, follows more or less the same principles, steps, and recommendations. Note however that the constraints are different, it is much easier to innovate to define new processes as effective as the original. In this regard, reference is made to the DMADV approach.

How to Succeed in the Lean Six Sigma Project

The success of the Lean Six Sigma project is directly dependent on the stated or hidden goals of the company approach.

Is it to improve the service delivered to the client so that all the members of the company benefit fully and durable or, more prosaically, the project has it any other purpose than to improve the profitability by all possible means?

Intelligent and cooperative management, coupled with an accompaniment of reasoned change, will be the right answer to successfully implement the first case. For the second, other specialists may have solutions.

How to Measure Just-in-Time Performance

Definition of Just in Time (JIT)

The market constraints needed for quite some years of reform production management methods in order to truly move to the notion of "pull flows." It is no longer planning that "drives" production but the market that "pulls" it. Long-term forecasting and on-the-ground production of standard products have been out of date for a few years now.

Just in Time

Traditional industrial production control systems are no longer adapted to the current needs of just-in-time and flexibility of production. They need to be supplemented with just-in-time systems. (Just in Time)

The rule is:

Produce tailor-made, according to the requirements of the customers respecting costs and quality.

This in-depth reform of the logic of production implies changes in the chain as much at the level of practices and methods of work as behaviors. The evolution of the relations between the actors of the chain of suppliers and subcontracting is the most obvious.

JAT is:

What it takes, when it is necessary, as it should be!

Kanban

Kanban is a production principle set up at Toyota factories. According to this principle, the Kanban makes it possible to trace the precise requirements inversely to the realization from the position Aval towards the post-Upstream. Note, Kanban is the name of the cardboard card accompanying the containers of parts being manufactured.

The Example of DELL (1)

Dell, the manufacturer of microcomputers, has taken the place we currently know, mainly through its integrated production management system, from order taking to the management of approvals.

Speed: Dell can deliver personal computers in less than a week, after telephone or Internet orders.

Reactivity: Since Dell manages very small component stocks (less than 11 days), it can immediately integrate and offer the latest technology.

Three Rules to Succeed

Empowering a maximum of actors for decision-making in the field where information is available, where the action is possible, and where risks are assessable. You want to have all hands on deck when you are deploying these types of systems. These people will be making prioritized decisions based on valuable risk management techniques as well as system rules.

Give out information rather than store it and keep it previously under the seal of secrecy. The free and generalized flow of information avoids having to deal with all malfunctions with expensive and devalued stocks, raw materials, and buffers. This also allows the company to have a full disclosure policy and transparency when it comes to the financial means of the company.

Measure performance precisely and continuously by not limiting yourself to the usual financial and productive factors, but in a way adapted to the needs and requirements of each decision-maker. See in particular the calculation of the TRS synthetic rate of return (*TRG Global Rate of Return*, TRG Global Rate of Return)

A Point on Dell's Strategic Model

Keeping in mind that Dell is an ever evolving company, and their standing within the technology, information, and financial realm may change at any moment, in the current day field, they have outlasted many competitors and continue to do so.

Leadership in a mutant market is rarely an assured position in the long run. The phenomenal development of the consumer notebook and mobile devices seems to have been right for a company-oriented strategy, and Dell is now multiplying its logistics strategic models according to their target market. Their main focuses are retail sales, custom-made orders, low-cost alternatives, and tailor made large accounts. They service both the private citizen and the large corporation.

MES Manufacturing Execution System is a set of tools positioning at the border between the too distant production management of the land and the technical supervision limited to driving. Integrated into the pyramid of the CIM, the MES supplies at the lowest level (shop floor) the essential functions for reactive piloting in the field.

Why the MES?

Other specific services are included in the concept. They are implemented on a case-by-case basis, such as Personnel Management, Document Management, Quality Management or Maintenance Management.

The MES communicates directly with the production management module of the ERP. It receives the production orders as well as all the work instructions (ranges, recipes, operating modes) and returns the production reports, the update of the production recipes and the elements of traceability.

The Four Main Contributions of the MES

•Optimal management of recipes and production ranges

•Real-time management of manufacturing batches

•Decision support available to operators in the field

•Specific products meet the first three points.

The fourth point, which is much more delicate, is also the most important. It is to assist the production manager in completing his daily concerns. The production manager is responsible for meeting goals and numbers set out, ultimately, by the consumer. In order to continue quality customer service, the production manager should have ample back up and assistance when needed.

How to Lead Production in Order to Deliver to the Customer on Time

When you are leading the production of a product, one of your main concerns will be creating the product, quality checking it, packaging, and shipping it in the allotted time stated to the customer. In order to be successful at this, as well as improve upon the processes there are several things that you will need to be aware of.

What are the components of such outstanding performances and what is the level of completion of the various lots? This is going to give you a baseline to know what you are working off of. If you find that you are not fully engaging in all components of an outstanding performance rating, then those things will need to be implemented. The same goes for the level of completion compared to similar departments both in and out of the company.

How to switch current productions on this or that unit so as to free resources for an urgent order? This is often one of those questions that companies don't answer until they find themselves faced with the task. You will want to have a fully fueled backup plan for when your resources are required in order to fill more urgent or rushed orders. If you do not have a plan in place, you will find you are abandoning one problem for another.

Mastering time cycles can be something that either takes practice, or incredible discipline. The management team within the department should have a full run down of each day, each goal, and each time frame to reach. These should then be broken down further by tasks. Make it as simple and step by step for those that are completing the process.

How do you, as a company, measure added value and cost? This can sometimes be difficult to calculate without a system like Six Sigma. Value and Cost are not always associated with a physical product. It can be time management related, opportunity cost, allocation of employee costs, etc.

It seems necessary to build a decision support system specific.

The definition of the main terms of the Six Sigma approach. A method necessarily relies on a very specific vocabulary lexicon. Each of the terms related to the method must be carefully specified. Six Sigma is no exception to the rule and has its own terms and acronyms. It is essential to know them well.

Six Sigma Terminology

As with many other programs on the market, Six Sigma utilizes both industry wide, and system specific terms. Within the system special terms are used to denote ranking within the system and the capabilities of that person to manage the project process using a Six Sigma background. The following list of terminology can be found on a regular basis within the system.

Black Belt

The "Black Belt" "BB": a step above the Green Belt. The Black Belt can take the project management and pilot several "Green Belts." The training is quite consistent.

Green Belt

The "Green Belt" "GB" is the first level of mastery 6 SIGMA. The Green Belt can play the role of team leader. This is the "active engine" in a SIX SIGMA project. He is relatively quickly formed.

Master Black Belt

The "Master Black Belt" "MBB" is a certification justifying a perfect mastery of 6 Sigma. The Black Belt Master must demonstrate successful Black Belt experiences and additional training. He is quite capable of leading a "6 Sigma" project throughout the company.

Capability

The capability of a process. Determines whether a process is able to meet the expectations of customer requests.

CTQ

Critical to Quality, critical quality parameters, critical to customer satisfaction.

DFSS

DFSS Design for Six Sigma. Defines a method that is particularly suitable for the SIX SIGMA project. A DFSS method includes a road map, specific tools, and a suitable training program. Its goal is, of course, to bring the company to the desired 6 Sigma quality level. It is based on the DMADV sequence.

DMADV

DMADV: Define, Measure, Analyze, Design, Verify.

DMAIC

DMAIC: Define, Measure, Analyze, Improve and Control. DMAIC is a method that can be considered a continuous improvement process in its own right. Based on statistical analysis, it aims to systematically eliminate all sources of non-quality.

DPMO

DPMO Defects per Million Opportunities. SIX SIGMA measuring unit. The DPMO indicates the number of defects per million units produced. The objective 6 SIGMA is not to exceed 3,4 DPMO or 3.4 defects per million units produced.

GIMSI

An overall design method of the performance measurement system. Gimsi is particularly suited to cooperative and continuous improvement processes.

SPC

SPC Statistical Process Control. Using the statistical tool to study the data produced to define the capability and the performance of the processes.

VOC

Voice of Customer, the voice of the customer

VOB

Voice of Business, the voice of the market

2

LEAN MEANS SPEED

Reducing Lean management to its French translation of "lean" is the main mistake of designers little aware of the complexity of a genuine process of progress. It is not a question of adopting an exclusively global vision centered on the reduction of costs and deadlines but of developing an approach from the field to the nearest real difficulties of the employees and collaborators of the company. But that is indeed much more difficult.

It is not difficult in an industry to do the necessary, but it is by making the superfluous that one earns money. Treat men as machines, and they make the necessary; treat them as men, perhaps you will get superfluous.

Definition of Lean Management

Lean management is a system of industrial organizations initiated in the Japanese factories of the Toyota Group (Toyota Production System), at the beginning of the '70s. It is a system of an organization more complex than what might be implied by the literal French translation, management "lean."

The objective is to improve the performance of the processes by exploiting the methods, techniques, and practices already available to the managers of industrial production. Just in time, quality at all levels of process, and cost reduction are on the agenda.

Viewed from a more practical, more concrete angle, the approach is essentially based on the active resolution of the recurrent problems of industrial production, whatever the field of activity. Therefore, the reduction of stocks, the fight against wastage and the reduction of faults, the just in time, the production with drawn flows (Kanban ...) and the control of the delays, the flexibility and the effective management of the competencies like the reduction of the costs, are integral parts of the process.

A Continuous Improvement Approach

If 6 Sigma is used to drastically reduce the variability of processes, lean, meanwhile, seeks not to reduce but to eliminate all that is unnecessary in the process, as the times waiting that impede fluidity, retouching that does not always commit to trying to do well the first time, overproductions that generate unnecessary stocks, unnecessary travel ... In short, everything that in theory can be described as wasteful and penalizes cycle times.

PDCA

Of course, this principle of reorganization is not considered as a new scheme to tackle once and for all the structures of the company. The whole philosophy of lean lies in the fundamental principle of continuous improvement. Progress is made step by step in a PDCA logic, referring to the wheel - Plan, Do, Check, Act.

The success of the project relies heavily on the ability to accurately measure progress against the performance objectives that have been set. In particular, read measurement recommendations for an industrial project.

The 7 Sources of Waste According to Taiichi Ohno

Taiichi Ohno is at the origin of the system "Toyota." He was adamant on waste reduction and understanding how that waste can negatively affect your company, your workers, and your bottom line. Waste Management, outside of your normal trash route, has been religiously implemented since Taiichi Ohno's explanation and implementation at Toyota. According to him, there are seven main sources of waste.

1) Enough Overproduction

Produce more than needed for the customer and create unsold stock to use resources or anticipate future orders. Take care to consider all parameters to properly size the batch.

2) Excessive stocks

Inventories are mostly the cache-misery of non-optimized production processes.

3) Defects, Alterations, And Rejects

Strive to do well the first time, treat procedures.

4) Waiting Time

An effective process is necessarily fluid. Attention to bottlenecks.

5) Unnecessary transport

Optimization of the locations of production sites.

6) Unnecessary displacements

Rationalization of movements and displacements.

7) Unnecessary treatments

Avoid unnecessary work for the created value.

And do not forget the eighth source of waste: the underutilization of skills. This is a dry loss for both the company and the employee. This is a major cause of de-motivation. This source of waste is not always the easiest to identify.

Measurement Is at The Heart of The Process

Measurement and performance indicators are the only ways to identify problems and ensure that you are on the road to progress. Without these, the company could be headed in a negative direction and no one would know until brick walls were smashed into. At that point in the process, especially for smaller companies, it is often too difficult to recover.

A Bit of Quality in Your Indicators

To put on its side some chances of success, it is particularly recommended not to be content to measure only quantitative quantities such as cost reduction and quantity produced. Accurate measurement of qualitative parameters, such as morale and willingness to participate in the company, will be the best guarantee for lasting success.

This aspect of the question is not always considered by the experts of the approach who are generally content to impose the solution and provide only vague sessions of information pompously called "accompaniment of change."

A Global Approach

Lean impacts all strata of the company and also requires full cooperation at all levels, the idea of lean management being to optimize the value chain over the long term. Lean

Management is therefore closely linked to the medium and long-term strategic approach.

The Limits of Lean

Before considering the implementation of Lean Management, it is better to study seriously the possibility of a truly cooperative approach. If we want a more successful outcome than the disastrous radicalizations dehumanized committed by the apprentice organizers, it is essential that everyone can, without any constraint, find its place and its role in the long term.

The Dictatorship of The Chronometer

The organizers who only reason in numbers, the clock in hand, and whose brain knows no other operation than the subtraction, apply a little too literally the principles of management lean the most radical. It is indeed necessary to focus on each step of the value chain, from the start to the final customer, in order to identify the wastes of all types (see Taiichi Ohno's recommendations above).

The Stubbornness of Wasting

However, all the waste, for an outside eye, is not necessarily one if we consider the process of value creation not in its PowerPoint representation, but in its reality: women, men and the time.

To over-seek to eliminate the "excess fat," the "lean specialists" and the direction obsessed by the economies quickly realized, drove straight a company to its loss: a rationalization of the processes too much Radical relayed also on the website of Les Echos.

The Principles of The Toyota Way Botched

Finally, we cannot develop a lean system, whatever the area

of intervention, without reference to the principles of Toyotism. They are fundamental.

Principles advocating a long-considered and consensus-based decision-making, which lays the foundation for a learning organization, are generally overlooked in the lean process on the ground. These are indeed the most delicate. They demand cultural reform and challenging traditional hierarchies. These two principles are nevertheless one of the main keys to the success of the project in the long term.

The Lean Six Sigma

Lean Six Sigma is, as its name indicates, is an approach combining the lean management method and the six-sigma method. The goal is to accumulate the benefits of both methods for optimal process improvement. Let's see the principle, project management, and its application in the world of services. Indeed, the approach is not only usable in the economic sector of industrial production, but also the cradle of the two original methods.

The Principle of the Lean Six Sigma Approach

How to do it right the first time? Without useless delay and without superfluous expenses. And of course, for the greatest satisfaction of the customers. All of these things can bring a company down in a heartbeat. You want to ensure that your company is not only meeting standards in these areas, but exceeding them as well.

Lean Six Sigma

Lean 6 Sigma is nothing more than a combination of two known process improvement methods known for its efficiency. The 6 Sigma method was born within the Motorola group; it is originally the fruit of drastic research to improve the quality of production processes. The Lean management method was developed in Toyota factories in the 1970s to improve turnaround times, introduce just-in-time and reduce costs.

Why Lean 6 Sigma?

Both methods, Lean and Six Sigma, are oriented towards customer perception. When implemented with caution, the benefits provided by the two approaches are fully compatible and complementary.

The activities that cause quality deficiencies in the customer's sense, as well as the delays that penalize processes, are some of the main sources of opportunities to improve quality, deadlines, cost, and profit share. Starting from this premise, the lean six sigma can then be considered as an essential step to improve customer service (according to the customer's voice), and overall profitability.

Management by Constraints?

A bottleneck may be characterized by an activity whose processing capacity is lower than the market demand. It slows down the whole process and seriously disrupts the service rendered to the customer. It, therefore, impacts overall profitability. In the industry, it will be a slower machine, an activity requiring a different pace, an inflexible supply chain or difficult to adjust. The process is unbalanced.

There is no need to dwell on non-bottlenecks. These do not penalize fluidity in the sense of market demand. It is better to take the time to identify and characterize the real bottlenecks in an economically viable dimension.

On the other hand, to imagine that one can solve all bottlenecks is a beginner's dream quickly forgotten from the first experience of industrial organization. It is, therefore, necessary to analyze, organize, delete and / or optimize the functioning of the bottlenecks. This is the fundamental of the theory of constraints. It's an iterative process. We will not solve the problem in a single pass.

Management by Constraints and Service Activities

This approach concerns all sectors, all types of processes. Look closely, take out your magnifying glass and your stopwatch / calendar. You will inevitably see the existence of bottlenecks whatever the sectors of activity.

The theory of constraints thus applies to optimize the processes of tertiary services but also the support processes of the information system, including ERP and the supply chain.

The Theory of Constraints, The Lean and The Six Sigma

The implementation of the theory of the constraints is not incompatible with the Lean Six Sigma. All of which is to articulate the principles and points of intervention. Many business owners dreamed then of "the factory without man," a factory with tense flow without any interruption. They were then in exactly the same state of mind as the industrialists and economists of the early twentieth century: The absorption capacity of industrial products is infinite and only the reduction of production costs matters. Economies of scale were, therefore, the only rule to follow.

The Integrated Factory

During the 1980s, the race for success necessarily involved improving productivity. In the spirit of the time, all the tasks of the production process were automatable and had to be automated. The systems implemented in this respect reached a level of complexity that made the whole complex and prohibited all forms of unforeseen events. Thus, these delicate maintenance systems have not been designed in a spirit of flexibility.

Flexibility

There was much talk, it is true at the same time, of the flexible workshop, but this was also controlled flexibility and planned in the design offices, far from the real constraints of the market of the 90s and ground. This vision where everything was predictable, programmable and automatable was not limited to the level of production. A global integration concept, the Computer Integrated Manufacturing (CIM), fueled conversations. All of the company's IT resources had to be connected to facilitate the overall automation of the plant.

CIM concept: Computer Integrated Manufacturing

The concept of integration and automation of the CIM was very much in line with the OST and its desires for extreme rationalization. For the OST, any job can be broken down into basic tasks. All that is needed is to automate all basic tasks and replace the man at all levels.

The Complexity of The Systems

The complexity of the systems realized in this spirit, induces many disadvantages. The automation and computerization projects have never been realized according to a real plan of global integration. The approach has always remained fragmented. One of the major difficulties today, is to communicate heterogeneous systems technically and functionally. We are once again facing an interface problem.

These systems also show a delicate development where complex systems are of course extremely long to conceive and very difficult to implement. They can also show a great fragility in both physical component and internal workings.

Often the notions of maintenance and degraded walking were not integrated into the system design. In fact, highly automated systems have often been shown to be not only fragile but delicate maintenance.

Rigidly limiting possibilities for flexibility can hurt a system. Complex systems are poorly adapted to the new market rules. In particular, involving a high speed of response to unforeseen and, above all, unpredictable situations.

Challenges of the Dashboard Project

Today, as we have seen, the stakes have changed. We are moving from the productivity stage to the quality stage and the controlled costs to the customer-oriented company.

We move from the concept that, "you will produce more for less of a cost," to that of, "you will satisfy the customer." This premise carries with it all the change in the vision of the company that we have already discussed during this study. The "reconfigured" company cannot do without man, pilot of complexity. Automation and computerization regain their nobility by being IN THE SERVICE of women and men.

Man can adapt to new situations. However, the ability to adapt does not only depend on past experience, but also on a breadth of view. It is even what distinguishes man from the machine. The machine works with more accuracy than the man but, on the contrary to this one, it is incapable of adapting to a different kind of work. You have to throw away the scrap and replace it with another one. But no one thinks of dismissing the workers because one has decided to adopt a new tool or a new method of work. (L. Chambounnd)

This remark, delivered here, anecdotally, dates from 1918.

In the industrial enterprise, the role of the field team is extended. To the actual conduct of the industrial equipment is added the need for monitoring and control of real-time production (flexibility requires).

If the activity of production is extended, that of decision hardly cuts there. In the unforeseen circumstances of the conduct (maintenance and safety in particular) are added the unforeseen of production. An analogy with the driving of a motor vehicle: it is no longer a question of only controlling the machine, it is also necessary to decide which way it will be desirable to borrow to go from A to B. Whatever the conditions, of course, it will be necessary to go to the end of the road in time and at the least cost.

CREATING COMPETITIVE ADVANTAGE WITH LEAN SIX SIGMA

M arketing can be characterized as a set of techniques and practices used to best identify the target of a specific product or range of products and thus facilitate its commercialization.

Based on a study of the expectations and motivations of actual and potential users, marketing is supposed to boost the commercialization of new products and services.

In other words, marketing is the way, rather than the art, to present the products or services in such a way that customers, real and potential, are persuaded to freely choose a product or service unavoidably.

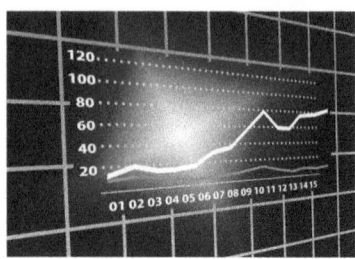

The marketing does not hesitate to draw its resources within the social sciences in order to succeed the transmutation of the superfluous into necessary, the useless into essential.

Marketing Strategy

Recycling: The art and the way of doing something new with old realistic goods. Marketing and its inseparable companion, advertising, are the darlings of modern business. Why spend in Research and Development, with the inevitable risk of failure when placing on the market, when it is enough to get back what we already have in store and whose profitability is perfectly controlled?

This is precisely the purpose of the marketing and advertising professions. Make new with old. Neither seen nor known. The risks are limited to the maximum. The examples are not lacking. They range from cars to cosmetics, yogurts, and pharmacopeia.

This is why, even in times of crisis, these two functions take up a large part of the budgets dedicated to development. Besides some bad language only see traces of innovation, vaunted by the apostles of the company, by visiting the marketing and advertising, as they are gifted in the art of recycling.

Management of Image

The brand is a signifier of added value and must be considered as a revenue generator. It is also necessary that its notoriety is up to the ambitions of the company. For this, we must build the brand, invest sustainably.

The establishment of a complete system of measurement of the adapted performance is, of course, necessary to ensure the relevance of the investments made. In terms of the brand, the measure must be particularly precise in order to understand the substance of each of the forms of values expressed by the brand.

Market Study

Presumably the best-known instrument of the profession. A market study consists of gathering as much information as possible in order to better understand the targeted target. The objective is to identify as much as possible a given market in order to be able to address it (or not) knowingly.

Market research is as much used to better develop an existing market to explore new markets. To be effective, the market research approach will be systematically associated with a battery of indicators as much productivity as effectiveness (definition of the problem, quality of the sample, the relevance of the target, consistency of the recommendations).

Segmentation

By dividing the populations, homogeneous according to the desired criteria, the segmentation ensures better targeting. Thus, it is then easier to adapt to the different actions, services, and products according to the segments of public identified.

Why not take advantage of the internet to push the cutting to the extreme in order to access the individualization of the offer?

That's what the "One to One" was already proposing quite a few years ago, perfectly described in the work of Don Peppers, Martha Rogers, and Bob Dorf, the One to One in practice. But be careful, in this case, we change the logic.

Choosing the right measures of performance is essential to

ensure the relevance of the breakdown and the quality of differentiation.

Scoring

Scoring is a technique of rating customer risk. In particular, banks use Scoring techniques to estimate the value of a client's default risk before granting credit. To segment customers more precisely, Marketing also uses adapted scoring techniques before launching a specific marketing campaign.

Customer Relationship Management

Permission Marketing

Rather than launching into an expensive marketing campaign where you try to reach everyone and anyone, why not contact the most interesting prospects and ask them for permission? This is Permission Marketing.

Dynamic Pricing

Dynamic Pricing is a dynamic pricing strategy based on market expectations and competing offers. The goal is to improve sales and ensure profitability.

Web Marketing

Web marketing is the techniques and methods to improve the performance of your website or online store. Web marketing covers many aspects of technology, analysis, and marketing.

Tribal Marketing and Viral Marketing

It's not new!

Until the late '70s, builders' automobiles could rely on a hard core of loyal buyers. Today, regardless of brands, vehicles of the same price range have ended up a little all look alike and have similar characteristics. With each renewal of the vehicle, the buyer now studies the offer, without giving priority to the brand of the vehicle previously owned.

Marketing and Sociology

By offering a completely differentiated offer, Apple Corp. was able to revive this feeling of identification with the brand as a sign of recognition of members of the same group. Marketing experts do not hesitate to use the term "tribe" to describe this behavior. It is also the fruit of a long work of putting into practice sociology of groups at the service of marketing. This is what is known today as tribal marketing.

Viral Marketing Definition

Viral Marketing is nothing but the word-of-mouth principle that we all practice without thinking, "Ah, I wanted to tell you, I bought a something really good..." This principle is now transposed on the net; it is called" Buzz Marketing." The idea is relatively simple. Ask yourself and ask yourself the following question. You will easily understand the principle.

The question: What is the best "media" to make a product known and lead to the act of purchase?

Put yourself in the buyer's shoes now, and choose your answer from the following options:

A radio ad or TV ad?

"No," will be the answer from the majority of buyers.

A page in a newspaper or magazine, without any targeting?

"No," on this one as well.

Maybe then the advice supported by a seller? To give him any confidence, one must be sure that he does not have a personal interest, a sales commission for example.

Finally, the advice of a friend or a user, in other words, word-of-mouth? Yes, that's for sure, like the vast majority of potential buyers, we are all more attentive to the advice of a loved one, and we listen carefully.

For the record, marketing is based on a study of the expectations and motivations of actual and potential users; it is supposed to boost the commercialization of new products and services.

A Practical Example: Starbuck And Lions Gate

A few years ago, Starbuck Company had an agreement with the Lions Gate film studios to discreetly promote the company's flagship films. Thus, in any case for the establishments located in the Anglo-Saxon countries, when the or the waiter approached you on the question of the cinema, it was not a girl or a nice guy who knotted contact to share his passion but well a pure operation of Buzz Marketing.

Viral Marketing and The Web

Many studies have demonstrated for a long time, that during a purchase it was easier to influence his choice to listen to the advice of a friend or a neighbor than to read the characteristics and tests performed by specialized journals.

The rise of the Internet among the general public has multiplied the phenomenon to the point where we are now ready to listen to the advice of complete strangers. As everyone knows, the online viral marketing system tends to drift. Many comments are manipulated, and the standard AFNOR on the treatment of comments online is hard to even reduce the number of false opinions and their adverse effects.

Web Marketing

Web marketing is one of the main neologisms used to designate the exploitation of the best marketing and sales techniques dedicated to the web channel, in order to better sell its products and its services on the Internet.

Improving the performance of a website or an online store is now part of the corporate strategy. Web marketing cannot be improvised.

Web Marketing Applications

The practice of web marketing has no other purpose than to improve the profitability of a website. Everything starts with the design of the site.

The Design of the Website

The central point of the activity. For the website to be sustainable, the design must excel according to the following aspects:

Aesthetics

The site is aesthetically appealing according to the audience of the moment. It makes you want to stay a little longer, come back and talk about it.

Ergonomic

The basic rules of ergonomics are respected. The new visitor is guided in his search and finds without waiting what he seeks.

Editorial

The editorial is particularly neat. Writing for the web is specific. The information needs to be accessible simple and quick.

Optimize indexing

It is essential that the website is sufficiently optimized to appear at the top of the search engine results for the main queries concerning the products and services offered. Rare are the visitors who exceed their expectations by the first suggestion proposed by the search engines.

Limit the bounce rate

Keep your visitors and limit the bounce rate. Attracting visitors is only the first step. Limiting the bounce rate, in order to turn them into potential customers, is the ultimate goal of the site. The bounce rate makes it possible to evaluate the number of Internet users who do not visit more than one page. We can always say to console each other that they have found at first glance what they were looking for. When this rate starts to be consistent, it is, however, preferable to intelligently resume the design of the site, the identification of keywords carriers and indexing in search engines (SEO Search Engine Optimization).

Website

Ensuring its presence and fame on social networks, professional communities, forums, groups, blogs, and Twitter will grow the client base. All these techniques are particularly expensive and time-consuming. The implementation of a measurement system to adopt a dynamic of controlled progress is fundamental: axes of improvement, objectives, indicators, actions, retroactive loop, and ROI calculation.

What is Marketing? One to One?

Getting to know your customers better is the basis of loyalty. Thus, we can better adjust the products and services according to their expectations.

By establishing a real dialogue between the marketing and the commercial on the one hand and the customer as an individual, on the other hand, the company understands better the needs of each customer and can thus offer a service or a product perfectly adapted to explicit expectations, even implicit in the best case.

The One to One Principle

Obviously, we do not personalize the relationship with all customers. This privileged relationship is reserved for high-potential customers, as well as all those who can be assumed to become profitable in a timely manner. So, we begin by calculating the "Lifetime Value" of each customer, to identify those who are eligible for Personalized Marketing, the One to One.

Lifetime Value Definition: A measure of the sum of expected net profits over the life of the client.

The Essential Ranking of Customers

An example of ranking the company's customers in four categories in order to identify deserving customers a privileged treatment. The goal of the company is to sustainably increase its profits. It would be as pointless as absurd to make efforts for less profitable customers. If we carefully calculate the benefit / cost ratio per customer, we discover many surprises sometimes. Customers who seem interesting are much more expensive than they report. They often solicit commercial services with projects that are not successful, still have problems, overload after-sales services, etc.

The Four Categories of the Customer

1. More profitable customers

This first column categorizes the most important customers profitable, that is, the minority that generates the largest share of the company's overall turnover.

1. Profitable customers

This second column groups profitable customers. They have significant potential value and deserve to be studied closely in order to apply the techniques of "cross-selling" or "upselling", in order to significantly increase the sales turnover generated for each one of them.

1. The customers in the making

This third column materializes unprofitable customers, but one can suppose that in the end, they will be able to become it. The potential value of each of them deserves to be studied on a case by case basis.

1. Unprofitable customers

Finally, this last column brings together unprofitable customers. It is better to take care of it as little as possible while respecting the contractual conditions. It is not at all obvious to get rid of an uneconomic customer. This is the solution to better guide the budget envelopes dedicated to marketing and sales.

Important note: the notion of "customer value" is inseparable from the notion of "value for the customer." If the customer does not find his account, he will choose another provider. Customers are not attached to the company. For the customer, it is now very easy to compare offers and choose the best of the moment. It is therefore prudent, as far as possible, to assess customer satisfaction.

A Clarification on The Notion of Personalization of The Offer

It is the role of the marketing and sales manager to know his client well in order to understand his explicit and implicit expectations and to serve them as best as possible.

Dynamic Pricing

Dynamic pricing is a strategy of dynamic price fixing, based on multiple parameters, in order to achieve an optimal flow of products and services according to market demand, while guaranteeing the best profitability. "

The best-optimized Dynamic Pricing algorithms take into account not only demand but also supply by closely scrutinizing price strategies practiced by the competition. Thus, airlines are practicing "Yield Management" to ensure the filling and the optimal profitability of its flights. I have understood well that all the passengers do not necessarily have the same motivations to make a trip. The executive in urgent displacement, forced to buy his ticket at the last minute, is not quite the same traveler as the tourist who is planning his trip for several weeks, they may be placed next to each other but have not paid their ticket on the same price. The principle of Dynamic Pricing then spread to many other market sectors, such as hotels and e-commerce.

The Principle of Dynamic Pricing

We all learned in primary the rule of setting the price of a product or service: The selling price is equal to the cost price plus a profit margin.

This rule does not always apply in these terms. Sales at a loss, which are allowed in many cases, are very useful for selling unsold goods and preparing the physical or virtual "showcase" of a range of appeal products.

Then the notion of the margin must be flexible. We do not always find customers for the price offered. Dynamic Pricing pushes this reasoning even further. It is about understanding the variations of the demand, that is to say somewhere the customer's psychology and the urgency of his need to adapt the price accordingly.

Defining the Selling Price: A Delicate Operation

It is indeed a question of finding the right price which on the one hand corresponds to the value which the customer has of the product and the service, and on the other hand, generates a sufficient profit margin. Of course, with the exception of some exceptional niche markets, we must also closely follow the pricing policy of the main competitors offering similar or replacement products or services.

A Price Depending on The Use and The Moment

The concept of a fair price is not universal. Each potential buyer judges the relevance of the price / service rendered. This estimate of the correctness of the price also depends on the expected usage as well as the moment. It is therefore obvious that the price catalog cannot be written in marble. It's about being flexible.

For example, the website Profitero.com, an observer of the e-commerce market, noted in 2013 that the site amazon.com made 2.5 million price changes per day. Walmart, "the little player," only made 55,000 price changes in a month.

As an anecdotal, as early as the year 2000, Amazon.com offered prices at the "head of the customer." The site offered for some products a price lower than the newcomer than the one offered to loyal customers and already well known (an article of Libération dating from November 6, 2000).

Bidding

In addition to this article on Dynamic Pricing, it is recommended to consider also the case where the price is not necessarily set by the seller.

The customer also has the right to offer his own estimate of the perceived value / purchase price ratio. This is the traditional and ancestral principle of "auctions," an essential instrument to achieve price equilibrium, so satisfactory for each of the two parties. The trader sets his floor price, and customers estimate an acceptable purchase price.

Classic Auctions

Bids start from a floor price, and interested buyers offer their bids to their personal limits. The highest bid wins.

Dutch or inverted auctions

Auctions start from a maximum amount and decrease until they find a buyer.

What is Permission Marketing?

Once the marketing campaign is ready to be launched, the targeted prospects are contacted personally to obtain their agreement. Only prospects who have accepted the offer will receive the presentation of products or services being promoted; flyers, videos, demonstration, and a proposal of tests.

The prospects contacted are of course much less numerous than with the conventional approach. But they are much better targeted and more receptive to commercials. The effort is less dispersed, and we can give all the attention to the right targets in a relationship of trust.

This is the foundation of a sustainable business relationship. If Marketing permission may not be too suitable for promoting consumer products, it is recommended for all higher end solutions. Permission Marketing is also the starting point for a good Viral Marketing or Tribal Marketing operation.

Ending Interrupt Marketing

Traditional marketing and advertising techniques have not changed much over time. It is a question of thoroughly exploiting the communication media: poster campaigns, mailings, advertising inserts in magazines, television or radio spots and the Web! It is marketing "pushed to the blind," intrusive and disturbing. We talk about interruption marketing.

The commercials seek to capture your attention and interrupt you in your tasks or activities of relaxation and leisure. These campaigns are very expensive for limited profitability. Marketing permission is a solution to intelligently exploit the web, no longer in a spirit of attention, but rather through mutual trust between the consumer and the promoter of the product or service.

The Principle of Permission Marketing

Permission marketing takes place in 4 major stages.

Stage 1. Get permission from the prospect

To engage the relationship and interest the prospect, it is recommended to offer a tempting offer. But this is not the typical product of appeal: "Ah you come too late! We sold everything like hotcakes". No. The offer must be truthful. Let's not forget that this is a relationship of trust.

Stage 2. Provide the expected information

As soon as the prospect grants his permission, he waits for the promised information. This is not to wait and enjoy his active and positive listening to clearly detail the offer. It is prudent to maintain motivation. To avoid the prospect forgets you and you cannot turn into a customer, it is necessary to establish a dialogue that cannot be summarized to a succession of commercials or reminders. This is a difficult step; permission is not always easy to transform into an act of purchase.

Stage 3. Expanding the field of permission

The business needs customers in particular. Prospects are only interesting once they have decided to take the "buy" action. We must continue and deepen the relationship in order to obtain a new agreement authorizing the communication of personal information, the willingness to receive samples if necessary and to test the product being promoted.

Stage 4. Make the sale

It's time to reap the fruits of the process and to go to the act of sale. If it is realized, the relationship is official. It will be maintained constantly remembering the rule too often forgotten:

Acquiring a new customer is still much more expensive than maintaining an existing relationship.

Beware, customers are fickle, and you are not the only one to practice active marketing.

Opt-in Opt-out

In order to prospect by e-mail, marketing exploits two very different modes of operation:

Opt-in: the user gives his authorization to receive e-mails.

Opt-out: e-mails are sent systematically. It is up to the user to ask for the stop of the mails.

The second mode of promotion is closer to spam than the smart marketing campaign.

Two Typical Errors

Error 1: Adding Leads for Nothing

Attention when choosing performance indicators. It's not about finding the largest number of prospects without ever transforming them. The goal of the approach is to increase the number of good, loyal and profitable customers.

Error 2: Attention to the Second Step

It is not a question of behaving like the seller of vacuum cleaners who is being opened, then puts the foot in the door to imperatively get the sale. It's about establishing a relationship of trust where the customer is happy to buy. That said, the seasoned salesman, knowing intuitively the technique of the foot in the door described by Robert-Vincent Joule and Jean-Léon Beauvois, knows that once the customer agrees to give more information, the sale is almost completed.

4

INITIATION: GETTING COMMITMENT FROM TOP MANAGEMENT

Definition of Process-Oriented Management
BPM, Business Process Management, or Business Process Management is a process focused on business processes. the company, customer oriented. The radical transition from a functional organization of organization to a radically transversal structure is particularly relevant to identify sources of efficiency and effectiveness and thus to guide improvement actions.

This file presents the main tools essential to the process: a definition of BPM, business processes, critical activities, process mapping, modeling and workflow in practice.

With BPM, business processes are at the center of a global process-centric process. Some clarification about the conduct of the project, it is indeed of strategic significance.

Business Activity Monitoring

BAM, an acronym for Business Activity Monitoring, translated into French by Supervision of business processes and activities. The BAM provides real-time management of the most critical business activities and processes. This chapter could also be entitled, the good use of the tools of Business Intelligence in the service of the control of the processes.

Business Processes and Activities

A business process is a list or culmination of assignments that ultimately leads to a completed product or service for a company. Business processes can also be used in order to successfully complete a company goal or organizational wish list. The business process itself is comprised of multiple, thoroughly described measurements that, when completed, result in one specific output. All of the inputs are listed beneath one of three categories including management process, business support, and operations.

During your business process, the project team will use an array of tools to keep the project timeline in check. One of the most important of these tools in the process mapping function. Process mapping is simply a visual look at the inputs and specific tasks that are to be completed in order to finish the project. Most organizations use process mapping software, as well as specific process maps. Some of the process maps include process charts, flowcharts, models, flow diagrams, and workflow charts and diagrams.

Business Cartography is also a way of X-raying the company, by revealing only the processes, their roles (in a customer-oriented dimension), their interfaces, and their interactions. This map of an organization will allow the project team to better utilize the processes they use when conducting in depth project assessments and creating timelines for those projects.

Process Modeling

Process Modeling is a popular and widely used way to visually represent a company's business processes and workflows. Through these models, usually implemented in graphs and charts, the company is able to find their weak spots, and work on those specific areas for improvement. Business Process Modeling covers two areas of a company's process, the capacity in which the company is currently functioning (As-is), and the capacity in which the company could be functioning with changes implemented (To-be).

Workflow

Principle Workflow, administrative workflow, procedural, or ad hoc. Workflow is the automation of information so that it follows predefined courses perfectly adapted to a company's production and exploitation.

Projects are not limited to the implementation of a specialized software package, as powerful as it may be. The BPM project is a strategic project. When completed, the BPM will reform the traditional operating principles of the company. The definition of a strategy centered process and not functional allows, when it is well conducted, to ensure a complete revision of the cycles of said processes.

And this, without rediscovering the reengineering. BPM Business Process Management is not the BPR Business Process Reengineering.

Agility

It is also preferable, in any case conceptually, to associate the BPM with the term flexibility or even agility. In absolute terms, the BPM makes it possible to reconfigure the processes without too much upheaval to adapt to the new situation. This is an essential notion that must find its place especially in the calculation of the ROI of the project. To follow then. But with the usual precautions required.

Human-Centric

Anyway, as for any complex project, the BPM project implies a close communication between the specialists of the technologies and the managers in charge of the business processes. In fact, a successful BPM is not just process-centric, which would not be sustainable. It must also be Human-centric. This is the ultimate target of the project.

The Metric

The definition of a precise and adapted metric is of course essential. On a technical level, a BPM project is in my opinion quite handicapped if it is not accompanied by a complete re-design of the functional applications by adopting an SOA Service Oriented Architecture approach (see also the EAI Enterprise Application Integration). On the manager side, the BPM is materialized by management dashboards and KPIs for Key Performance Indicators (BAM Business Activity Monitoring).

The Principle of BAM Business Activity Monitoring

These include approaching real time. Exploiting the integration and inter-exchange of intra and inter-company production applications, the BAM seeks to offer the best perception of the performance of activities and processes. From a technical point of view, the BAM exploits the concepts of application integration and communication, the EAI and the related concepts such as SOA and integrates into a more global dimension of BPM.

From the point of view of the user, responsible for activities or processes, the BAM solution is embodied in management dashboards composed of KPIs for Key Performance Indicators, refreshed in "real time." We are in a process-oriented perspective of global application integration in the more general context of performance improvement.

Definition of a Process

ISO 9000: 2000 defines the process as a, "Set of correlated or interactive activities that transforms input elements into output elements."

This definition is actually quite incomplete. Let us consider the general definition proposed by Le Petit Robert.

"A process is a set of phenomena conceived as active and organized in time."

This second definition dates back to 1865, and I do not think the authors were already considering the needs of BPM. It specifies, however, three essential notions: a set of active phenomena, organized and expressed in a temporal dimension.

The definition of the dictionary lacks only notions of finality and measurement to become exploitable in the 21st century.

An example of a process explained during the online training.

Process offers a cross-business-oriented, non-linear, and function-oriented vision. The transversal vision is neither more nor less that of the customer. It is by adopting this type of representation, dynamic and not static, that it will be possible to perceive and thus improve customer-oriented cycles.

Activity

An activity, within the scope of an organization, is an identifiable task with clearly defined inputs and outputs whose added value is measurable. The equation to represent an activity within a company would be where I=Input, O= Output, and V= the Valued Added. Ultimately the equation would read, I+O=V, where the input and output are given structured numbers to represent their worth within the process project.

zation's dynamic analysis process. Whether as part of a targeted improvement process or a more global reorganization, process modeling allows you to formalize the precise functioning of an organization using a standard and easily understandable language.

The semantic richness, offered by the techniques and tools of organizational modeling of the company, thus facilitates a common perception of business processes oriented "improvement" punctual or continuous.

The quality of the realization of the modeling phase of the business processes like that of the establishment of the cartography of the processes besides, condition the relevance of the image that we mentally create the operating principles of the company. This phase deserves all the attention required to avoid any kind of misinterpretation and thus miss potential competitive advantages.

Modeling Methods

UML, Unified Modeling Language, is an object-oriented modeling language. This language is the result of a combination of the three object-oriented methods: Booch, Object Modeling Technique, and Object-Oriented Software Engineering. The three experts combined their experiences and skills to define a standard and universal modeling language. UML is supported by the OMG Object Management Group. See also on the website UML Project Manager.

BPML

Business Process Modeling Language

XPDL

XPDL, the XML Process Definition Language, is offered as a standard process description language. It is supported by the Workflow Management Coalition (WfMC). XPDL is built from Extensible Markup Language. XPDL defines a standard exchange format, in particular, to facilitate the use of graphics tools or description of processes of different origins.

BPM

Business Process Modeling

BPMN

Business Process Modeling Notation is a standard graphical representation of business processes.

The goal is to use a common language to facilitate the realization and communication of modeling. The BPMN project initiated by BPMI, the Business Process Management Initiative, is supported by the OMG, Object Management Group.

OSSAD Method

OSSAD, Office Support Systems Analysis and Design, is a method of analysis, design and implementation of information systems developed as part of an ESPRIT project. This process modeling approach is at odds with its older MERISE or SADT who focused almost exclusively on technical aspects, offering a description of the operation and behavior of IT systems. To know a little more, read the reference book cited below or download the OSSAD method on the website of Philippe Dumas, co-author of "The OSSAD method" Organization Editions.

Workflow Within BPM

The workflow enables the modeling of business processes in the framework of BPM, Business Process Management, to have a more global approach.

Administrative Workflow

To better understand workflow on the administrative level you could look at the following:

The management of loan requests for a banking organization, claims processing files for an insurance company or the reimbursement of expense claims for any organization type are application examples for administrative or tertiary workflow tools.

For each of the above cases, the documents to be developed follow a specific path that includes verification, investigation, information or approval tasks. IT workflow tools make it possible to define the path of each type of document, to control the deadlines and to precisely follow the execution. The workflow tool is sufficiently flexible to allow the pre-defined circuit to be modified during the course of the trip in order to avoid any form of blocking.

Ad-hoc Workflow and Procedural Workflow

The ad-hoc workflow, unlike the more locked-down procedural workflow, allows users to dynamically intervene on the path in order to deal more effectively with unforeseen events. The workflow is not a recent concept. The idea of automating workflows is as old as the generalization of enterprise computing in the late 1970s. Since the concept has evolved well as computer tools, standardization, and modeling.

The workflow applied to the tertiary sector is an instrument of rationalization quite similar to the already old practices practiced in the industry.

For the record, the workflow is also not a recent invention related to the deployment of IT. During the 1920s, many administrations used internal pneumatic tube circuits to transport documents.

5

SELECTING THE RIGHT PEOPLE— AND THE RIGHT PROJECTS

I f the management of skills, predictive or not, is limited to the function sheets, the HRD completely misses the hidden skills. Hidden skills are the talents that everyone develops but do not appear in any procedure or flow chart. Yet without the hidden skills, so unclassified and unidentified, the business cannot work.

HRD: Predictive management of hidden skills competencies

Faced with the shortage of skills available on the market for many trades, it is better to know the capital "skills" of the company.

Logically, each skill must be used, as much according to the interest of the company as that of the employee. These two objectives are not necessarily concomitant. Ineluctably, divergences emerge over time. These differences sometimes result in a break, especially if the ambitions of the employee are frustrated by the requirements of the company.

For HRD, the match between the expectations of the company and the aspirations of each employee is a concern at all times. It is in this light that postings to new posts and training campaigns will be defined.

The Ideal HRD

In an ideal company, at least in the sense of the management of women and men, HRD is able to accurately measure employees' skills potential, their know-how and their skills. Knowing the precise needs of the company, present and future, it can propose the best role, competence and adequacy. Always in this ideal case, HRD is also able to identify future needs and trigger, at the right time, recruitment campaigns, anticipate the obsolescence of trades, and offer the appropriate training.

It will then be able to evaluate for each present and future need of the company, the opportunity to recruit or start training campaigns to compensate for the obsolescence of available skills.

HRD in Real Life

But in real life, it's not so simple. In particular, it would still be necessary for HRD to be aware of the company's main strategic lines. Provided, of course, that the company itself has built a strong strategic line.

Short-Termism

Cost management should also not be forgotten, but in any case, put in the background, so as not to deprive itself of real talent because of a simple accounting vision short term. Economies without vision, short-termism, are the plague of business survival.

Hidden Skills

To continue in this direction, it is extremely difficult to identify, and formalize, what might be called hidden skills, those that do not appear on the record, and without which companies do not or could not work in spite of what the streamlining rationalizes think.

These include:

- The ability to communicate, that is to say, the faculty of employees to exchange and help each other beyond well-designed organizational charts.
- The sense of responsibility and the taste for training the latest arrived, while nothing has been planned to put the foot in the stirrup of the new hired.
- The ability to develop relationships beyond the sphere specified by the function sheets with other professionals, internally and externally. It is an essential skill, but impossible to formalize.
- And finally, the magic talent, that many employees have, to find good ideas and unlock seemingly inextricable situations.

Pay attention to your skills management. All those who do not understand the importance of hidden skills, and they are very numerous, wreak havoc during restructuring.

Dismissal Policy

Having a thorough, legal, and fair dismissal policy is extremely important within an organization. While the company wants to reserve the right to dismiss employees for unsatisfactory production, the employees also want to feel secure and protected within their jobs. Ultimately, for a company, they are required to have a company policy that reflects the core ideals of the company but meshes with state and local laws determining fair practices. It is best to include legal input when creating or changing this policy. Measure Public Performance

Public services are also forced into the race for performance. On this point, the turn has been taken in recent years as we can already see it every day.

The quest for efficiency to provide taxpaying citizens with decent quality service without blowing budgets is perfectly legitimate. It is however particularly damaging to systematically associate the notion of performance with that of financial profit.

The Purposes of Public Service

Public Service and market sector do not have the same ends. Applying the rules of market logic to the public sector can only be concluded by the bursting of the latter. Thus, the rationalization of resources and means must not only be justified by the economy without a precise assessment of the degradation of services rendered. Public service is not a business.

Performance of Public Service

Public Performance It is therefore important to choose the terms that are used for performance. Value creation is not just synonymous with increased profit. Value creation can also be understood as an improvement of services for citizen satisfaction (1). In any case, this is to understand the formula when we approach the theme of public services.

In any case, the quest to improve the diversity and quality of services offered to all citizens without distinction is nothing more than the putting into practice of the republican principle (read also on a subject related and topical The perversions of the resulting culture).

Value Creation and Public Service

Managing public performance on a macroeconomic level, at

the state level and over time, the creation of value thus heard is clearly more profitable in every sense of the term.

John Kenneth Galbraith, an economist advisor to Roosevelt and Kennedy, drawing on the experience of the United Kingdom at the end of the war, noted the importance in terms of the values of progress and the quality of public services, rather than posting a substantial GDP. This remark is all the more relevant in these current times when "sustainable development" is essential in all debates.

Without affecting the satisfaction of public service employees, this is self-evident. Professional conscience is the foundation for increasing the quality of services rendered. But there can be no professional awareness without a minimum of comfort during the exercise of his office. Employee satisfaction is, therefore, a prerequisite.

To go further, read more Definition of the *LOLF Organic Law on the Laws of Finance.*

What is LOLF?

This law globalizing the management of the state budget has been in force since 2006. In principle, the LOLF aims to reform state management to facilitate measurement of performance, decision making, and accountability. The LOLF proposes a budget no longer divided in kind of expenditure but into public policies. Missions are supposed to bring more transparency to the citizen and better control of parliament.

CAF Public Service

Self-Assessment Framework CAF, the Public Service Self-Assessment Framework, is an instrument intended for the management of public bodies by applying, according to the designers, the principles of total quality. The CAF is in line with the Deming PDCA cycle and proposes a self-evaluation with reference to an "Excellence" model structured according to 9 main criteria and 28 sub-criteria. As the reader knows, the CAF is inspired by the model proposed by the EFQM.

Critical Comments

The reorganization of public administrations has not escaped the formalists always in search of "excellence." Personally, I am not sure that the improvement of the performance can be summed up in a distribution of satisfaction and bad points in reference to a standardized model and theoretically "perfect" according to its designers. All this seems very presumptuous (read also, *The Perversions of the Culture of the Result*).

Both standard setters and evaluators are by definition far removed from realities on the ground. This is true for the industry; it is also true for services, and this is no less true for public administrations.

Christophe Dejours, who holds the CNAM Chair in Occupational Psychology, has devoted much of his work to studying the impossible gap between the work prescribed by managers and the actual work done by the actors on the ground. To refer in particular to the short but rich work: The evaluation of the work to the test of reality.

On the other hand, it seems to discern to me in half-tone behind this quest "of excellence" of vain inclinations having no other the object of perpetuating the archaic model of Dad's organization, hierarchical and authoritarian. Improving performance, whether in a competitive context or not, is a bit more subtle than that.

So, I will not spend more time on this theme. To know a little more about the CAF model and the criteria of "excellence" selected, I invite you to refer to the EIPA website of the

European Institute of Public Administration which describes it precisely.

The Public Performance Dashboard

The budget reform formalized by the LOLF, Organic Law on Finance Laws, aims at an in-depth modification of the way of designing the management of public services, public organizations, state administrations, and local authorities. The design and implementation framework proposed by the Gimsi method will serve as a concrete guide to the implementation of the management dashboard for public administrations.

The 5-step Design Method

A Dashboard can get started using a structured five step design method that will quickly and simply guide your employees through the process. Let's quickly take up the 5 key points of designing a dashboard.

Know Your Audience

Whether the creation of this performance dashboard is for public servants, public organization, administrations, or authorities, it is vital that you understand their specific needs and wants. The dashboard will not be the same across all sectors of both public and corporate organizations. Always pinpoint the exact sector before you begin the creation.

Select the Objectives

Then, to concretize the action in the field, there is no other solution than to define clear and precise "tactical" objectives. They are the product of previously defined areas of progress. The actions will be channeled and targeted to achieve the goals as best as possible. In order to do this, it is important to ask a lot of questions. It sounds so simple, but too often we don't ask enough, or the right questions. These answers will shape the entire project.

Choosing Indicators

It is essential to have a set of relevant indicators to measure the effort effectively, to ensure that the objectives are achieved in time and at the expected cost, to monitor, evaluate and refocus the actions undertaken. The indicator selection phase is at this stage of the project.

It is highly recommended to reform practices and customs in practice in organizations and especially in public administration. Indeed, it is no longer a question of seeking an exhaustive measurement by collecting excessive amounts of indicators of all kinds. It is necessary to change logic and move to the control of the performance.

To be fully effective, controlled management is content with a reduced number of performance indicators KPI, Key Performance Indicator, less than ten in general. Efficiency indicators oriented along the lines of progress defined above will be favored at the expense of conventional productivity indicators. They will not be left behind. They will be selected sparingly in order to counterbalance and balance the previously selected piloting indicators (efficiency).

Collection of Data

There are technological tools to ensure a significant part of the technical aspect of this complex part of the project. Grouped under the generic term ETL (Extract Transform and Load), they constitute the Business Intelligence project.

But beware it is risky to summarize this step to the technical operations of cleaning, formatting, and consolidation of data as delicate as they may be. Information is closely linked to power. The saying is always true. Especially when dealing with budgetary aspects (execution and monitoring).

In order to increase the chances of collecting "good information," especially when one is encroaching on the prerogatives of those less concerned by the project, it is better to seriously prepare for change.

Composing the User Interface

This portion of the process is usually handed off to specialists within the field. They will take all of your organized findings and information and ergonomically design a user friendly interface. This section of the dashboard is essentially for the implementation once delivered. After all, what's the point of creating a process if it can't be understood well enough to be used?

Public performance is a topical issue if ever there is one. For some, public services, in essence, should not be confronted with the logic of performance. For others, there is no reason for public services to remain outside the logic of markets. This is a snapshot of the situation.

It is true that as a citizen, user and taxpayer, we all expect quality service, without discernment of situation or location, in accordance with the basic efficiency rules. Is this the case for applying the logic of commercial enterprises, the purpose of which is not even profit, but rather the continuous increase of this profit? In my opinion, it is well in the acceptation that one lends to the term of performance that the question arises.

6

PREDICTING AND IMPROVING TEAM PERFORMANCE

Balanced Scorecard

The Balanced Scorecard is not a new type of dashboard to measure other axes than the only financial perspective. Although this is one of the components of the BSC, it is not at this level that we must look for the originality of the method.

All the major players who have looked at pilotage issues and scorecards over the last 25 years have naturally come to the same conclusion.

Why the BSC

On the contrary, the Balanced Scorecard proposes a new mode of management and management of the company by relying on the establishment of a rigorous framework for the elaboration and deployment of the strategy guaranteed by the permanent balance of the 4 perspectives. This is to highlight and master the cause-and-effect links.

This is where the true key of the method lies. The evaluation of the performance is carried out using scorecards, and the remuneration of the managers is directly linked to the performance.

The Balanced Scorecard does not graft like a patch supposed to disconnect the company from its bad one's habits. The BSC spirit starts as soon as the strategy is developed and continues throughout its deployment. Without a real mobilization of all the moments, the project is not realizable.

The Principle

The designers of the BSC recommend considering the company according to 4 very specific perspectives. The performance objectives, the performance indicators, and all the corresponding actions fit into the dimension of each of the four perspectives.

Financial Perspective

It is not only a question of measuring financial performance but also of ensuring the efficient use of financial resources.

Customer Perspective

The perception of performance but from the customer's point of view. With time and experience, this perspective also extends to the key stakeholders of the company, i.e., those in the narrowest circle.

Perspective Organizational Learning

Performance from the perspective of human capital, information system, culture, etc.

Before embarking on a Balanced Scorecard approach, it is important to understand that everything begins with a reform of the way of thinking, of designing the strategy and of piloting. This reform must be chosen and not suffered. Otherwise, it is failure assured. We will flee like the plague solutions that do not integrate the importance of this true revolution.

The Issue of Dashboards

The theme design tends to occupy a prominent place quite representative of the unfulfilled expectations of companies in terms of pilotage.

Until the last decades, the issue of piloting assistance was indeed less present. When the context was a little more stable and the competition a little less systematic, to seek the continuous increase of the productivity, as well as the decrease of the costs of return, was still the best of the strategies. The scorecards of that time, limited to exclusively economic and productivity measures, were entirely adapted.

For already two or three decades, the context has changed significantly. To guarantee the real profitability of the capital invested, it is necessary to elaborate more substantial strategies. Following only the financial measures is not enough. The loop is too slow and does not react in time.

Therefore, continuous progress needs to be monitored along the lines chosen by the management team when developing the strategy. You will need to ask how the company envisions it's competitive advantage. Do they view it by their ever diminishing delays? By their unwavering customer service?

Or by a constant renewal of products and offers? Different companies will have different answers to these questions. Whatever the chosen competitive advantage, once the strategy has been developed and implemented on the ground, each of the axes concerned must be measured precisely to ensure real and continuous progress.

Measurement and Strategy

Kaplan and Norton are also not the precursors to establishing a direct link between activity measurement and strategy.

The basis of the dashboard theory is based on this assumption. If not, what good could the measure be for if it is not to drive according to the chosen paths of progress?

It is true that the dashboards were in fact directly related to budgetary requirements and were exclusively local and tactical.

This era is now outdated, and next-generation dashboards measure performance across all pathways of progress defined by the strategy. This design approach is not unique to the Balanced Scorecard.

The Balanced Scorecard does not add up as a patch that is supposed to get rid of the company's bad habits. The Balanced Scorecard spirit starts as soon as the strategy is developed and continues throughout its deployment. Without a real mobilization of all the moments, the project is not realizable.

Measuring Performance

In a highly disrupted and competitive environment, it is important to measure performance in all its aspects and not just measure financial performance. The financial performance is indeed a particularly slow loop whose effects are only perceptible in the longer term. The exclusivity of the measurement of the financial performance does not allow to perceive in time the essential signals suggesting the imminence of correction of trajectory, a change of course or a more radical reorientation.

The 4 Perspectives of the Balanced Scorecard

For Robert Kaplan and David Norton, the success of the implementation of the strategy, i.e., when the creation of values is at the rendezvous, depends on the quality of the management according to four perspectives very specific performance.

With the Balanced Scorecard, Robert Kaplan and David Norton discuss how to decline performance according to the following sectors of business..

- Financial Perspective
- Shareholder Performance
- Customer Perspective
- Company Performance
- Internal Processes
- Internal Advantages
- Growth
- Forward Momentum
- "Balanced" Dashboards

The important word in the balanced scorecard is "balanced." It is vital that there is a balance between the short and medium to long term objectives. That there is balance between financial and non-financial indicators. That there is balance between external perception and internal performance. Balance within the company as a whole can have a stabilizing affect that moves across the board.

Declining Performance

For many years, companies have simply measured performance by tracking financial results. This vision, which today is described as partial, was nevertheless sufficient when the context was stable, and the speed of evolution was relatively slow.

In a context of rapid change and exacerbated competition, performance in all its forms is highly recommended in order to make the best decisions and not just measure financial performance. This is indeed a long-term loop that does not allow to act in time.

For Robert Kaplan and David Norton, the successful implementation of the strategy, that is to say when the creation of values is at the rendezvous, depends on the quality of the management of well-defined perspectives.

The 4 Balanced Scorecard Perspectives

With the balanced scorecard, Robert Kaplan and David Norton propose to decline the performance according to 4 perspectives:

Financial Perspective: What to bring to the shareholders?

Customer Perspective: What needs to be brought to customers?

Internal process: What processes are essential to the satisfaction of shareholders and customers?

Organizational learning: how to manage change and organization.

In its first delivery, the Balanced scorecard proposed an approach based on 4 essential points of support. Transformation of the strategic vision into operational objectives allowed for the opening of doors for improvement. Communication of the strategic vision and objects to all within the company, as well as, customers, would allow the company to link performance grades to specific departments or even employees. From there planning from the ground floor up forced them to decline the objectives set and set quantitative goals to accomplish. From here they would exploit the feedback received, exploit the looping, and learn new strategic operations that they would later implement.

Warning

This method is much more difficult to implement than it seems.

Recall that according to the authors, 60% of organizations do not create any link between the budget and the strategy. Either way, 85% of executives spend less than an hour a month talking strategy.

This is enough to begin to consider the extent of the reform of mentalities and the difficulties to overcome to bring it to an end. It will also be prudent to be wary of the freely inspired Balanced Scorecard approaches proposed both in Europe and in the United States by consulting firms specializing in problem simplification.

But as for any major project, the customer will not waste his time deepening the question by immersing himself in the heart of texts and books here selected. Knowing your subject well, both in terms of needs and possible solutions, is still the best way to judge the "capability" of a provider.

Strategy Maps

A Strategy Map, a term that should be translated into French as a strategic map preference map or a strategic roadmap, is a cause-and-effect diagram presenting the relationships between the different strategic objectives according to the 4 perspectives (financial, customer, process, learning, and growth).

Strategy Map is the keystone of the Balanced Scorecard project framework. It allows to "materialize" the passage of the expression of the strategy to the creation of values proper.

Main Gains

According to the authors, the Strategy Map, a strategy map, is an indispensable tool to:

1) Clarify the strategy and facilitate the communication of this strategy to each employee

2) Identify the key processes for the success of the strategic implementation

3) "Align" the human, technological and organizational investments so that they work in the direction of the strategy

4) To highlight the differences of implementation of the strategy and thus to facilitate its correction

Recommendations

It is strongly recommended that the Strategy Map fits on one page.

Remember that the financial perspective is not necessarily expressed in the same unit of time as the other performance axes. This loop is longer term. It is, in fact, the result of the other 3 perspectives.

The success of strategic objectives is directly dependent on people, culture and management, in short, organizational quality. Remember that the success of strategic objectives is also strongly linked to the fluidity of information (SI).

If the Strategy Map is a tool for concrete expression of the strategy, it is also an instrument of information within the company.

Balanced scorecard or dashboard?

Balanced Scorecard

The Balanced Scorecard approach proposes developing the strategy respecting the balance in 4 perspectives: Financial, Customer, Internal Process, Organizational Learning. They are complementary. To appreciate the chain of value creation as a whole, the four perspectives are to be considered, see also critical success factors and strategy maps.

Balanced Scorecard Financial Perspective

- What value is created for shareholders?

This is the purpose of the capitalist enterprise. For this first perspective, the concept of value is nothing other than the financial return of the company to shareholders, its sole owners.

Customer Perspective

- What value is created for customers?

But to serve a return to owner-shareholders, you still need to generate profits. For the most general case of commercial exchange, the customer expects products or services of quality corresponding to his needs and sold at an acceptable price.

Perspective Internal Processes

- What is the performance of the key processes of success?

A business is a set of processes. To satisfy the customers, the processes are optimized to ensure not only the traditional triptych: Quality (intrinsic), time and costs but also innovation and quality of services.

Organizational Learning Perspective

- What is our capacity to progress?

A good company is a sustainable business. For that, it is essential that it progresses and that it is renewed. In other words, its durability is closely linked to the quality of management that will (or not) motivate employees and offer them opportunities to evolve not in a personal capacity but in the service of the company.

The Balanced Scorecard concept is a typical example of a structuring approach to build a perfect capitalist company where all resources are optimized and leveraged by increased shareholder value. This is also why the approach is essentially vertical, top - down.

Strategy Maps

Strategy Maps Strategy Maps are the central point of the system.

They are the expression of the strategic hypotheses and define the cause-and-effect relationships between the results measures selected and the determinants of performance.

"Each measure selected for the Balanced Scorecard must be part of a chain of cause-and-effect relationships expressing the strategic direction of the business (Robert Kaplan, David Norton, the Balanced Scorecard Organizational Editions).

The establishment of the Strategy Map requires a lot more work than necessary. The quality of the management system is directly dependent on the relevance and likelihood of the Strategy Map.

The 4 Perspectives of the BSC's Explained

The balanced scorecards now allow to better frame the

conception of the strategy and above all, its deployment, the true Achilles heel of the strategic approach. The 4 perspectives presented above are still the fundamental pillar of the process.

The objective of any strategy is to ensure a long-term satisfactory return on capital employed. Financial indicators, which are oriented towards profitability, such as Return on Investment, EBITDA, EBITDA, or EVA, are used to evaluate the performance of actions undertaken in the past.

Perspective Internal Processes

What processes deserve our "care" at all times to satisfy customers and shareholders.

The quality of the services delivered to the customers is directly dependent on the performance of the processes. It is important to identify the key processes that can improve the supply and, consequently, the profitability of the shareholders.

This category encompasses all processes that contribute closely to the creation of values without omitting the longer cycle processes such as those related to innovation.

Perspective Organizational Learning

=

With the last two books balanced scorecard strategy maps and Strategic Alignment, the two authors have slightly changed this formulation: How to "align the intangibles" that are people, systems, and culture to improve the critical processes?

To achieve long-term goals, it is essential to renovate the infrastructure. This axis concerns three chapters: men, systems, and procedures. Progress to be measured focuses on training men to access new skills, improving the information system and matching procedures and practices.

Balance is not an empty word.

The important word is "balance." The French translation in "prospective scorecards" does not highlight this essential characteristic. It is better to use the term "Balanced Scorecard" which, though still incomplete, is still closer to the original design mind. The balance of the 4 perspectives is indeed essential. One must never penalize one axis for favoring another, but on the contrary highlight the causal links of the 4 perspectives. This is how, according to the authors Norton and Kaplan, the return on invested capital will be effective.

The Balanced Scorecard Approach

One begins to understand that it will not be enough to add dashboards here and there with indicators drawn right-to-left, measuring any of these axes, or in total the 4, saying "that's it! We do Balanced Scorecard!" It is important not to stick to this unique teaching. The "Nirvana of the ultimate method" is not so easy to access. By the way, talking about the importance of non-financial indicators is becoming the new cream pie of niche market consultants.

A Fundamental Reform

The implementation of specialized tools is only the last step and not the body of the project. It is also the easiest step. Because we must not forget that beforehand, we must develop a concrete and realistic strategy, in line with the model defined by the two authors Robert Kaplan and David Norton.

It is a fundamental reform of the minds of the management and all the actors of the company that must be undertaken.

The establishment of the Strategy Map (highlighting causal links) and its application are far more complex than the set of tools. It requires other skills.

It is not a homeopathic approach that must be adopted, but rather restorative surgery. The operation is heavy and the side effects on the operating modes of the company more than consistent.

The Balanced Scorecard, it's not something that we add like this, thinking that tomorrow, with the new tool, everything will be really better! Hello, "headaches" when you try to apply them within companies, SMEs or not, with their own designs of the strategy, effective for the rest!

For the whole of the method to be operational, it is necessary to follow the process of "conversion of the spirits" from end to end.

A tip: Book the method only for companies that do not develop a strategy (they are many) and are ready to accept the principles of well-square and standardized (there, they are much fewer.

Attention: The project is long and very expensive. When preparing the budget estimate, we must also consider the time spent by all executives to integrate this new way of thinking, to convince employees and to polarize actions.

Advantages and Disadvantages of Balanced Scorecard

The Balanced Scorecard approach has many positive points, and it encourages managers to better understand the many aspects of performance.

Financial indicators are thus balanced against customer-oriented indicators, processes and growth dynamics. It is not a question of unduly favoring a type of performance to the detriment of other axes. The integration of the dynamic perspective of growth is also one of the strong points of the method.

However, the method can be criticized for the exclusivity of the Top Down approach. The implementation of Balanced Scorecard in companies tends to endorse pyramidal hierarchical structures. For many companies with little focus on people and organizational culture, Balanced Scorecard implementations are closer to implementing new extended control tools than to true performance measurement tools.

If one wishes to stay in the logic of the designers, it is good to remember that the Balanced Scorecard is originally oriented in general directions. To decline to the operational directorates to express all the substance of the process is already less simple than it seems.

As for deploying it to the greatest number, the bet is lost in advance.

If we still want to take advantage of the Balanced Scorecard framing, it will be to facilitate the development of the operational strategy based on the strategy maps while taking into account the limits of the latter. For the rest of the project, the Gimsi-type methods, focused on decision-making itself, are clearly preferable.

7

LEAN SIX SIGMA LOGISTICS

L *ogistics Supply Chain Management*

Originally, the term logistics belongs to the military world and thus designates the best management of supply, housing, and transport of troops.

Subsequently, the term has been extended to the world of business to designate the management of material and material flows, in and out, of stocks and transportation of products.

Supply Chain

Supply Chain is the Anglo-Saxon term used to designate the supply chain. This term corresponds more or less to that of logistics seen above. In practice, when you choose to use the term supply chain instead of the traditional term in France logistics, it is often in the spirit of Supply Chain Management, a much more complex concept. See below the definition of this last term.

Supply Chain Management

Supply Chain Management defines all resources, means,

methods, tools, and techniques intended to drive the global supply chain as efficiently as possible from the first supplier to the end customer.

It is, indeed, link after link, to estimate as accurately needs, availability and capacity, to better synchronize the elements of the global supply chain and manufacturing.

For large companies using many subcontractors, this is the only way to serve customers according to price, time and quality requirements. If a company embarks on an SCM project, it wants to improve flows and deadlines while ensuring rigorous cost control.

In fact, we must go even further to understand the concept. Supply Chain Management is not just a family of enterprise software products to facilitate the management of the chain, known as "SCM."

It is a question of anticipating needs and being able to deliver the right product, where it is needed, when it is necessary while ensuring optimal control of costs and quality. This concept of control goes far beyond the technical management of flows. It is indeed a matter of shaking up the ideas received in order to establish solid cooperation with all the partners of the chain in a spirit of shared competitive advantage.

Conceiving the Supply Chain

If the absolute quest for cost reduction has long been the main trigger for supply chain implementation projects, it is, however, inseparable from the requirements of regularity and flexibility of supplies and production flows.

To satisfy customers and maintain market share, the fluidity of supplies, a condition sine qua non for the continuity of end-to-end services, is unavoidable. With the company extended to multiple economic partners, the issue is now much more acute. The relevance of the implementation and the rigor of operation of the supply chain determines the overall viability.

The Core Business, Controlling Costs, and Processes

For already two good decades, companies are engaged in a policy of outsourcing on a large scale. Virtually all functions are now eligible for outsourcing.

This policy of refocusing on the core business meets a clear challenge of controlling costs. But it is important not to interfere with the overall effectiveness of the processes involved. Indeed, the quest for cost floor cannot be the only criterion of differentiation in a hyper-competitive dynamic like the one we are currently experiencing.

The fluidity and regularity of supplies and production flows are equally important.

To ensure continuity of service to the customer, the risk of losing control over each phase of the overall process must be eliminated. This is the sine qua noncondition for the viability of the supply chain.

To limit the losses of control on the whole process, until the last years, there were not really other solutions that the centralization. Only operations with low value added or not involved in customer processes could be outsourced.

Since the advent of a competitive climate exacerbated, the price of the value "customer" has soared. To capture and keep them, we must keep prices down, innovate and above all be very fast. The cooperation between all partners involved in the production process is the only solution to this equation with 3 unknowns.

With the development of information technologies such as

supply chain management and supply chain management, this cooperation is now possible. The information systems of the different partners are interconnected and communicate.

If, as a client, I can know the manufacturing capabilities of my supplier, and if my supplier can, in turn, know my present and future needs, we can both work at the fair, deliver quickly and well customers, control costs and ensure our respective margins. We are thus moving from a climate of general mistrust to widespread cooperation.

Principle of Supply Chain Management

Supply Chain Management or how to move from a jerky and static management of orders / deliveries / stocks, where ignorance and mistrust prevail, to dynamic and continuous management, based on extensive information exchange.

Definition of SCM in 3 points:

1. Supply Chain Management, SCM, defines the management of the supply chain from the first supplier to the end customer.

2. The objective of the SCM is to assess as accurately as possible the needs, availability, and capacities of each link in the logistics and manufacturing chain, in order to better synchronize them and serve customers in the best possible conditions.

3. Supply Chain Management improves flow and time while controlling costs.

The Challenges of Supply Chain Management

A product is never realized end-to-end by the same company. Many suppliers, intermediaries, and subcontractors are involved in different phases of product realization. When these companies do not communicate, supply can be broken at any time, and the flow of production is slowed down or even interrupted. Flow decreases and delays become longer.

To prevent the unavoidable hazards of production, there is hardly any other solution than to build security stocks, both as inputs and as outputs. These stocks will be more or less substantial depending on the case. In any case, intermediate stocks are a handicap in a current production context based on product customization and just-in-time.

To guard against businesses, each at the level of the procurement process, constitute buffer stocks of greater or lesser importance, which are costly in any case. These stocks not only limit the flexibility of the production, but they increase the costs of production.

The key: The exchange of information.

If no action is taken to change the operating principles, the communication between the different entities is limited to a minimum. They are different companies, and without forcing the trait, traditionally relations are reduced to the classic: order form, delivery, and invoice. Anticipations, as well as manufacturing concerns, are not exchanged. Hence the partial deliveries and stock outs. The slightest error slows down the flow or interrupts it, and the whole chain is penalized

By globalizing the vision of the production process, Supply Chain Management makes it possible to change logic. Being of a compartmentalized and product-oriented nature, the production becomes extensive and customer-centric. The supplier upstream of the chain is not concerned more than it is necessary to use the delivered products once they have been loaded into the truck and are out of the factory. The only information available to him is order forecasts and firm orders. He, therefore, organizes his production with this information.

In a context of tense and "tailor-made" flows, it is a question of considering the process in its totality to best serve the customers according to the three criteria of judgment: time, quality and costs. Logistics must be "managed" to size each phase of the process, and best serve the demand.

Supply Chain Strategy

What we call strategy is basically the art practiced by the Cook agency. It essentially consists of crossing rivers over bridges and crossing mountains through passes

Logistics, supply management, is no longer the poor relation of a self-sufficient industry. Far from it. The generalized outsourcing policy has rebuffed the cards.

The supply chain is now the backbone of the company. The strategic issues raised by this new configuration of the production chain go far beyond technical considerations alone.

Supply Chain Management is not limited to choosing the right SCM software product and using it. Supply Chain Management is in itself a decisive factor in gaining a competitive advantage.

Maintaining a low price is not the only argument that can attract customers. Speed and innovation are equally decisive. The main challenge of the chain will be precisely to solve this complex equation for several partners, for the benefit of the customer.

Only the search for a strong partnership of "mutual trust," based on real extended cooperation (win-win), will ensure the maintenance of the optimum of the equation: reduced price * fluidity * regularity * responsiveness.

The Logic of Trust

For this, it is still necessary to be part of a logic of "trust" and be ready to share as much operating data as forecasts or ambitions. It is indeed to avoid the effects of surprise so that each partner can anticipate.

Reactivity at the Heart of the Strategic Approach

Not only is the engine of fluidity, but also that of responsiveness, as in the case of the launch of a new product. The speed of placing on the market is often decisive. The quality of the responsiveness of the global chain will make a difference. In conclusion, the strategic approach of the Supply Chain is inseparable from a cooperative approach.

Measuring Supply Chain Performance

Supply chain performance conditions the success of the strategy. Whether you're looking to capture new market share, build customer loyalty, or improve the profitability of each customer, or all three at once, the supply chain plays a vital role. Measuring the performance of the supply chain, therefore, deserves special attention. Let's see all this.

Performance & Quality of the Supply Chain

Achieving a global quality level worthy of the name is, of course, the key factor to the viability and efficiency of the supply chain.

Continuous improvement projects will be cross-cutting, which is not easy. Cooperation and trust, as we saw in the development of the strategy, are also inseparable from the concept. Conditions for Achieving a Successful Supply Chain

The Supply Chain must be in perfect harmony with the company's strategy. The performance of the Supply Chain is a concrete condition for the feasibility of most committed strategies. In fact, it is always a question of either winning new market shares, or building customer loyalty, or improving the profitability of each customer, or all three at the same time. How to achieve this goal without a supply chain "set to the little onions"?

The Supply Chain must be aligned with customer needs. In fact, it is necessary to consider the entire process from the first supplier to the end customer. The idea being that in an optimal system, there must be no break in this global chain, no fragile link is acceptable.

Easier said than done. It is not always possible to guard against the failures of a supplier by searching for second sources. Some complex subsets require close collaboration with the subcontractor. It is not easy to change suppliers as easily. In all cases, the solution is anticipation as far as possible. And begins do we anticipate? By sharing information. This is the fundamental rule of Supply Chain Management.

The supply chain, when it is well designed, is a lever for growth or even an element of differentiation vis-à-vis the competition. Well-designed and, above all, well-managed: it is a question of precisely following the unavoidable objectives of costs, quality, and deadlines and of giving each of these fundamental parameters the importance required by the strategy followed.

Finally, the Supply Chain must be adaptive.

Customer needs are changing; markets are constantly changing. It's about being responsive at every link in the chain. A single angle of view is to be considered: the overall performance of the chain.

To control the performance, a well-designed dashboard is essential.

Traceability

Traceability, while quality, requires special care. Quickly, traceability is nothing but the reverse path from the finished product in the store or the customer to the exact identification of each of the elementary components used for its production, procedures, and means of production.

(Solutions: fine notification, marking, RFID ...)

- Production and Just in Time
- Here are some principles of management of the performance in industrial production
- Inventory Management Inventory
- Inventory
- Optimization Inventory
- Management Strategy
- Purchasing Management
- See here some principles of B2B Purchasing Management
- The lean and the Supply Chain

It is quite exceptional for a company to give the logistics department a role and a power to impose its will on other services. This is the source of major difficulties. It is mainly about negotiating with Engineering or the design office who is working on new improvements and new products. It's their role. They hope that the improvements will be integrated without delay and that the new products will be operational as quickly as possible, regardless of whether the subcontractor's schedules are already overloaded.

Just-in-time manufacturing is struggling to smooth out its demand, it is demanding changes in batch quantities and disrupts, that is to say, the manufacturing programs. Subcontractors are obliged to undergo not to become a bottleneck.

We must follow if we want to maintain its status as the main source of this client.

The purchasing department of the customer has no other logic than the reduction of unit prices. They always have good solutions to explain to the subcontractor how to reduce costs while mastering the quality.

In reality, if the client controls the tight flow and the zero stock, it is because the subcontractors keep them stock to cushion the backlash. It's a reality. It is even more exacerbated in a context of increasingly short series with many changes of ranges. "Lean" is a bit like that too.

Supply Chain Performance Indicators

Traditional performance indicators of the client: costs, quality, the delivery rate on time, order compliance is thwarted by suppliers who maintain inventory and significant sorting rates. How to make lean if the order is fluctuating? How to meet the requirements of the supplier? How not to cut our margins too much? The problem is the range changes.

Supply Chain Management Package

The choice of SCM Supply Chain Management product is of crucial importance. Well managed, it can itself become a lever of competitive advantage. It must facilitate the flow of information (Business Intelligence), anticipate, be flexible and adaptive so as not to hamper innovation without delay and interconnect partners' information systems. See the technical aspects of the "SCM package." See the offers of suppliers such as:

SAP Supply Chain

Oracle Supply Chain

Oracle Cloud Supply Chain

And for a complete analysis consult the "*Magic Quadrant*" of the Gartner Group, updated annually and presents the positioning of the SCM providers according to well-defined criteria.

Leading Supply Chain

The Supply Chain plays a strategic role in its own right. The performance of the complete supply chain will be evaluated taking into account the importance of taking a potential competitive advantage. In the same logic, the Supply Chain management model described here, as well as the indispensable dashboard of the responsible team, faithfully reflects these issues.

Supply Chain and Value creation

To ensure efficient management of the supply chain, it is good to keep in mind the notion of value creation. Representing the global value chain in the Michael Porter sense makes it easier to identify the value creation process.

Adjustment to the fairest levels of stocks and their distribution in order to maintain fluidity and regularity present and future (anticipation for a maximum of cases configuration of production needs, no break).Each entity, link in the supply chain, is independent. So, she built her own strategy. It is prudent to know this and take it into account when developing the project. It is not a question of satisfying only the requirements of the client. In many cases, these requirements are inconsistent with the expectations of the partners.

Synergy is a nice word but requires a lot of effort. However, the SCM cannot be optimal without this synergy. The traditional approach of relations with subcontractors according to a master / slave is still well established in the French industrial fabric. In this case, they are not partners, and the originator limits the sharing of information to a minimum.

Without thinking from the outset of its project in terms of creating value for all partners, it will be impossible to achieve the optimal system. Note, the cooperation must appear in the performance indicators. It is an essential measure.

The SCOR

SCOR Supply Chain Operations Reference model is a reference model for the implementation and management of the Supply Chain supported by the APICS Supply Chain Council (APICS). The SCOR model advocates themes. These themes include the Identified process such as plan, stock, produce, manage returns. The measure of the processes, tasks and activities, and the exploitation of good practice. Oftentimes the three levels of modeling (Strategic, Tactical, and Operational), is used. Formalization is, of course, the main advantage to using the levels of modeling.

ASLOG Benchmark

ASLOG, the French Supply Chain and Logistic Association, has defined a set of eight indicators that serve as a benchmark for establishing a benchmarking benchmark, in search of logistical best practices. These eight quantitative indicators are listed here:

- Forecasts rate
- Customer service rate
- Customer complaint rate
- Production service rate
- Supplier Service Rate
- Purchase Forecast Reliability Rate
- Global Logistics Cost Rate expressed as a percentage of Net Sales Turnover
- Inventory Rate
- The supply chain management project, mastering the supply chain

The only way to succeed in the supply chain management project is to look at it from the point of view of creating values for all partners concerned from the very first lines of the preliminary study. Reduce Supply Chain Management project at the sole supply management would be a mistake. The SCM project will be considered as the realization of a total organization system in its own right.

The success of the project is closely linked to the quality of the relationships established with the partners of the first circle of proximity.

The Supply Chain Management project is part of a broader approach to take into account the specific expectations of the partners. Well designed, the SCM must ensure flow management and complete control of the extended chain.

Strained Flows

In an ultra-competitive world, as we have known for three decades, the "client" is no longer a preserve. To find them, to attract them and then to keep them, there are few other solutions, with a few exceptions, than to keep prices down, to innovate constantly and to be very fast. All cycles have indeed accelerated, the cycles of production, delivery, design and marketing innovations. Without cooperation between all the partners, it is almost impossible to manage these cycles without breaking. The communication between the different entities, the exchange of essential information such as forecasts or availability, is the only solution to this equation with multiple unknowns.

Once the system is in place SCM, it is then quite possible to entrust to more efficient partners that it is for the question of the final cost, the quality level of the availability and especially of the speed. High value-added activities such as design are also concerned. Supply Chain Management will provide flow management and complete control of the extended chain.

Supply Chain Management and Synergy

The Supply Chain Management project is of strategic importance. To build an "extended" company, the traditional relations between suppliers and principals limited to the exchange of service information are no longer appropriate. It is now a matter of building a common strategy, at least for aspects related to the SCM theme. It is undeniable that the overall system including all partners is viable only under this condition.

Quality of the Exchanges

Information will be widely exchanged, shared, explained and documented. It is precisely the quality of the flow of information that will provide the necessary anticipation to put an end to slowdowns and supply disruptions penalizing the entire customer offer. It is also the wealth of information exchanged that will take the shift to flexibility in both production and design.

Reactivity of the Supply Chain

Reacting quickly and being able to set up the production line of new ranges within acceptable deadlines are also guarantors of customer durability and increased overall profitability.

Cooperation and Performance Measurement

The quality of cooperation cannot be decreed. It is earned by scrupulously applying the principles of a continuous improvement approach. The Supply Chain management dashboard will necessarily include the appropriate performance measurement indicators to closely monitor progress.

The type of inventory and supply management practiced internally and by the most direct partners determines the performance of the supply chain. screen. This management therefore significantly affects the quality of the service provided to the customer.

Stock and Just in Time

In a just-in-time context, when production and supply centers, just like customers, are spread across the territory or internationally, we can no longer rely exclusively on traditional methods warehouse and store management (ABC, 20/80, Wilson's method ... see below).

Inventory Management

The "dynamic" management of material flows, semi-finished products and finished products now makes up for "static" inventory management.

The transition from a "static" management type of stocks to a "dynamic" management type of flows, will not be done without reforming the organizational modes and operating principles of the supplier-customer chain since the first element of the chain until the end customer. Customer-oriented logistics management is indeed another level of complexity. It involves close collaboration with all partners in the logistics chain.

Confidence and Transparency for Dynamic Logistics Management

The systematic sharing of relevant information, anticipation, trust and development and the implementation of common strategies are the main characteristics of the indispensable change in supplier-customer relationships for management. inventory adapted to the current needs of just in time.

Reminder: Inventory management methods, classical techniques

Method ABC and 20/80

Methods ABC and 20/80 are directly derived from the work of Vilfredo Pareto.

20% of the same product references are found in 80% of orders. Or, 20% of product references generate 80% of turnover. It is therefore essential to ensure that these precise references are never out of stock and are as accessible as possible. The ABC method makes it possible to identify them and to divide them into 3 classes.

The Wilson Method

The Wilson Method is a formula for estimating the inventory values of warehouses and stores taking into account demand and cost of ownership.

Why do we need to optimize inventory and app management?

Abolition of stocks has the effect of directly sending the pressure of demand directly to the workshop Luc Boltanski and Eva Chiapello. The new spirit of capitalism stock management to justify the importance of stock control; it was still customary a few years ago to stick to the only capital costs.

Even when the issue of expiry does not come into play, good management requires that the time between the purchase of a product (money out) and sale (money input) is as short possible, regardless of payment deadlines.

Stocks are equally penalizing on two other aspects. They hide the malfunction of flows and discard the requirements of essential optimizations of the global strategy (the thresholds chosen for minimum stocks and "security" in particular). They are a major obstacle to rapid renewal of ranges and more generally to innovation in practice.

Storage and Flow Management

Flow management is an inexhaustible source of questions whose complexity, well known to mathematicians, constantly tests logisticians.

The use of specifically dedicated simulation software makes it possible to test beforehand the opportunity and the performance of a new configuration. Taking into account the fundamental limitations of modeling, of course.

Modeling Software

Improving its management of material and product flows is an essential point to ensure perfect optimization of its cash flow, WCR, working capital requirements, and cash flow. It is important for companies managing stocks of materials and / or finished products to adopt the principles of a continuous improvement approach and to set up an effective management system.

8

DESIGN FOR LEAN SIX SIGMA

Six Sigma is a process improvement methodology that is born from theories such as the Total Quality System (TQM) and Statistical Process Control (SPC), it also incorporates many elements of the Deming PDCA cycle. At the beginning of 1982, this methodology began to be popularized when the engineer Bill Smith introduced it in Motorola as a business strategy to improve the quality of the products, later it was popularized and improved by General Electric (GE). The Six Sigma concept has transcended beyond these companies and has become a new administrative philosophy widely disseminated worldwide.

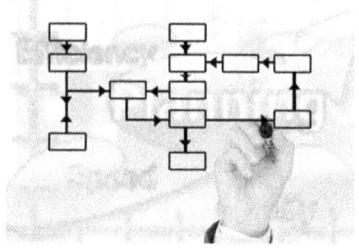

Sigma is a Greek letter whose main base is the standard deviation. Six Sigma has the focus of reducing variability and defects in products or processes. The goal of Six Sigma speaking from the technical aspect is to reach 3.4 defects per million (DPMO) understood as a defect to any product or part that does not meet the requirements established by the client. This corresponds to fluctuation between +6 sigma and -6 sigma of the average value.

The experiences of the companies that have decided to implement Six Sigma can indicate from global figures of reductions of 90% of the cycle time or 15 billion dollars of savings in 11 years (Motorola), productivity increases of 6% in two years (Allied Signal), up to the most recent of between 750 and 1000 million dollars of savings in a year (General Electric).

Lean Manufacturing Integration Model - Six Sigma with principles of Concurrent Engineering adaptable to SMEs

The application of the Six Sigma is characterized by the implementation of 5 concrete stages (DMAIC): Define, Measure, Analyze, Improve and Control.

Define - Here you create the definition of the project where the problems are included and opportunities, the objective of the project, the expected benefits, what are the things that are within the scope of being able to improve, the structure of the team and the timeline of the project. The next step is to define what the client's requirements are in order to establish the metrics with which the results will be measured; these requirements are called CTQ's (critical to quality). Some questions that must be resolved in this area are the following:

• What processes exist in your area?

• What activities (processes) are you responsible for?

- Who or who owns these processes?

- What people interact in the process, directly and indirectly?

- Who could be part of a team to change the process?

- Do you currently have information about the process?

- What kind of information do you have?

- Which processes have a higher priority to improve?

Measure - The first step is to be sure of having good metrics through validation or analysis of system measurements. All factors affecting performance are defined using methods such as Pareto, cause and effect diagrams, cause and effect matrices, failure modes and effects, and process mapping. The CTQs are used to determine the indicators and types of defects that will be used during the project, and the current results are compared with the client's requirements to determine the magnitude of the improvement that is required. Some questions that must be answered:

- Do you know who your customers are?

- Do you know the needs of your customers?

- Do you know what is critical for your client, derived from your process?

- How is the process developed?

- What are the steps?

- What kind of steps make up the process?

- What are the measurement parameters of the process and how are they related to the client's needs?

- Why are those the parameters?

- How do you get the information?

• How accurate or accurate is your measurement system?

Analyze - In this stage, graphic analyzes are made to see the relationships and impact that they have the factors that are going to be measured with the outputs that they report (CTQ's). A series of statistical analyses are made as hypothesis tests, confidence intervals or design of experiments to see the significance of the factors that are measured (inputs). Subsequently, opportunities for improvement are visualized, according to their importance to the client and identify and validate their causes of variation. The questions that must be asked are the following:

• What are the customer's specifications for their measurement parameters?

• How does the current process perform with respect to those parameters? Show the data.

• What are the objectives of the improvement of the process?

• How did you define them?

• What are the possible sources of variation of the process? Show what and what they are.

• Which of these sources of variation do you control, and which do not?

• From the sources of variation, it controls How do you control them and what is the method to document them?

• Do you monitor sources of variation that you do not control?

Improve - First potential solutions are identified through team meetings and Brainstorm or use of TRIZ. It is very important to have a complete measurement system to be able to correctly validate the improvements later; this validation can be done through the Design of Experiments, once these steps have been completed a detailed project plan must be completed with cost analysis. benefit. Here is the answer to the following:

• Are the sources of variation dependent on a provider? If so, what are they?

• Who is the provider?

• What are you doing to monitor and / or control them?

• What is the relationship between the measurement parameters and the critical variables?

• Do critical variables interact?

• How did you define it? Show the data

• What adjustments to the variables are necessary to optimize the process?

• How did you define them? Show the data

Control. - The control strategy is determined based on the current process map, modes of failure and its effects and a detailed control plan. Next, the controls identified above are implemented. After determining the capacity of the process with all the improvement and control of it, finally, the process must be monitored.

The questions to solve are:

• How accurate or accurate is your measurement system?

• How much has the process been improved after the changes?

• How do you keep the changes?

• How do you monitor the processes?

• How much time or money have you saved with the changes?

• How are you documenting it? Show the data

The Six Sigma methodology will initially do the following:

• Identify each activity or process

• Know and understand the company's clients

• Distinguish key processes that add value to customers

• For each key process, obtain specific client requirements through surveys and analysis

• Convert customer requirements into useful measures (CTQ)

• Measure each metric (CTQ) and for each one establish a normal distribution graph

• Set the limits of each metric and the number of defects that fall outside these limits

And then he will repeatedly:

• Identify processes and metrics (CTQ)

• Do research work

• Improve the project team and document all the steps

• Obtain process measures

• Analyze the measurements and look at the causes of the defects

• Generate solutions, implement improvements and measure changes

Among the tools used in Six Sigma are the following:

• CIP, Processes of Continuous Improvement

- Design / Redesign of Processes

- Analysis of Variance, ANOVA

- Balanced Scorecard, BSC

- The Voice of the Customer, VOC

- Creative thinking

- Design of Experiments, DoE

- Process Management

- Statistical Process Control, SPC

- Response Surfaces

Six Sigma helps to improve existing processes but recently when introducing new products or services. Design for Six Sigma is a more generalized term used to introduce new products and services that meet customer requirements at the time of its launch. It is a design process that requires a total understanding of the process steps, capabilities and performance measurement.

The main objective of Design for Six Sigma is to attack the vulnerabilities of the design from the conceptual to the operational part by deriving and integrating tools and methods for its reduction or elimination.

Contrary to the DMAIC methodology, the steps of Design for Six Sigma are not universally defined. Almost always companies use the Six Sigma Design version that their service provider provides them. The expected sigma level for a product or design using DFSS is at least 4.5 sigma but may be higher depending on the design used.

The design for Six Sigma requires great emphasis on:

- A scheme that represents the voice of the consumer

• Define all CTQs starting from the consumer's point of view

• Visualize all the problems that the project may cause and technical failures

• Project management in such a way that all parties involved can communicate efficiently

• A detailed project of changes in the processes

There are 4 phases of which consists of the DFSS: Identify, Characterize, optimize and Verify

Identify. - Designing the objective and scope is the most important part. A program of Project can be used to simplify and clearly note what will be designed. The design team should write all the information they may need, in particular, the voice of the customer (VOC) and the voice of the company (VOB). With the help of the QFD, the definition of the requirements that will be grouped later in systems and sub-systems (components) will be achieved. One question that designers must consider is where the project will go, how much it will cost, if they have the necessary resources if it will be easy to operate and maintain and the methods by which the service, maintenance, and transportation will be carried out.

Characterize. - The design team must carry out a number of solutions. Is important that you write or draw each idea on a piece of paper, which will help you remember and describe the ideas more clearly, it will also be beneficial to discuss it with other team members if the drawings exist. Once all these ideas are carried out, the team must decide on one. To be able to do this, tools such as Morphological Matrix and Pugh Matrix can be used, where all the functions and their possible solution parameters are shown.

Optimize. - Once the preliminary design is available, optimization is carried out, which can be deterministic or statistical. In the statistical part, optimization must be carried out in such a way that the design is not affected by external factors called "noise factors" that may affect its performance. The transfer function should be used to deal with this step. Detailed documentation of the optimized solution must be made, which must include all the information needed to produce the product or service.

Verify. - In this stage, the team can make a model or prototype. The respective tests and evaluations to see if it will work properly as planned and comply with the requirements defined above. It should prepare the places where the product can be produced for its launch. At this point, you must be sure that the product has a market and that no competitor can beat them.

Concurrent Engineering (CE)

The product development cycle begins with the conception of a need based on market analysis and research and development activities. Commonly, a series of steps are used to design the product, identify the processes through which it will pass, machine the parts, assemble the components and send the products to the market. The product designers focus on the functionality of the product and its performance leaving aside the process by which the product or manufacturing limitations will pass. However, the decisions when making the design are of great importance and should be taken into account as they will directly impact the cost of the product, its manufacturing possibilities, its performance, the time it enters the market and its quality.

Concurrent Engineering is a philosophy and not a technology. Gladman was the first to introduce the term Design for Production in 1968, which emphasized that the products had to be designed in the first instance for production, which would make the manufacturing resources effectively used to obtain the maximum benefits. The Department of Defense of the United States paid attention to the concurrent Engineering, promoted its study and implemented it in the weapons production system.

Some technological terms focused on the process have been used to describe the functional concepts of Concurrent Engineering. Among them is Design for Manufacturing (DFM), Design for Productivity, Design for Testing, Design for Reliability, Design for Installation, Design for Good Service and Design for Assembly (DFA).

There are two main fields where the CI can be implemented, by work teams or by Software. The formal way to implement is through work teams, where the team consists of designers and individuals related to all the areas described above, which must be chosen for their ability to contribute to the design of the product and processes that are capable of identify problems early and implement actions to solve them from the design to avoid the great cost that it would take to start the product again after moving on to another stage.

Estimation of Costs

In Concurrent Engineering it is very important to try to estimate the cost of the product in the design phase, in such a way that the gain that is sought from the product can be achieved. If the projected cost of the product exceeds the cost limit considered, it should be considered not to continue with the development of the product or redesign it.

The cost of the product must be estimated from the costs of material, machining, testing, assembly, transportation and all tasks related to it. The estimate of costs must be made from the first stages of the design since at this stage it is where the cost of the product is directly impacted.

Technology

Computer systems are used in Concurrent Engineering as a support to carry out tasks through software, which acquires, represents, integrates and coordinates the knowledge required for Concurrent Engineering. There are several programs for this function as tools through CAD CAM CAE, where today you can interact with product design, assembly, logistics machining, etc. This is very helpful when several departments interact with the prototype and can make changes together.

Knowledge Flow

The knowledge that is generated when designing or making a product should be stored in a database for future reference or as an aid when required to make changes to the product or design a new one. Many managers and engineers have difficulties in finding information for the multiple databases that come to exist for a single product, for this type of situation it is advisable to use the concept of Unique Data Base (SDB) that includes all the design information, analysis, process, planning, tools, quality, etc.

The implementation of Concurrent Engineering can be divided into 3 phases:

Getting Started. - Which begins by recognizing the desire to improve the process of product development and ends when the company decides to investigate / implement the Concurrent Engineering.

Planning / Preparation. - Includes information and analysis and ends when a plan Complete Concurrent Engineering is presented and approved.

Make. -starts with the first change of the selected process, this part never ends.

Some common organizational barriers are:

• Lack of support from top management.

• Unsuitable work climates.

• Protective functional managers.

• Inadequate reward systems.

• Lack of consumer involvement.

• Lack of involvement by the provider.

• Fear of losing creativity.

Design for Manufacturing and Assembly DFM / A

The design for Manufacturing and Assembly is of great importance in Concurrent Engineering as it helps to reduce the time it takes to develop a product by avoiding design errors, and parts of the product that are difficult to machine.

The assembly is almost always the most intensive part of work and the one that contributes greatly to the size of the total cost of the product. Automated robot manufacturing has been used lately in the industry to assemble products in very large quantities. However, not all products are recommended to put this type of manufacturing by robots. With the design for assembly, it is sought that all the parts with which a product has been designed in such a way that its assembly is facilitated, and therefore its cost is reduced significantly.

The Design for Assembly (DFA) was developed at the University of Massachusetts, where they were based on 2 main factors: The ease of handling and assembling the components and the number of parts used in the product. This tool is a methodology of a systematic approach for the improvement of designs, used to simplify the structure of the product and reduce manufacturing costs.

Among the qualities that stand out are the following:

• Product simplification

• Competitive market tool

• Estimation of cost and assembly time

• DFM / A integration for the total cost of the product

• Product simplification

• Negotiation and communication tool with suppliers

• Selection of process and material

• Early estimator of costs.

The actions that should be used most are:

• Minimize the quantity of components

• Use commercially available standard components

• Try to remove screws

• Use common parts through product lines

• Design to facilitate the manufacture of parts

• Design in such a way that no mistakes are made during assembly

• Use modular design

• Form parts and products to facilitate packaging

• Delete or reduce the required setting

In order to carry out the DFM / A, a company must make changes in its organizational structure, in such a way that it provides a closer interaction and better communication between the design and manufacturing personnel.

Design for Reliability (DFR)

Reliability can be defined in several ways:

• The idea that something fulfills the purposes with respect to a certain time

• The ability of a device to perform as designed

• The resistance to failure of the device or system

• The ability of a device or system to perform a function on specific conditions for a period of time

• The probability that it will perform adequately during a certain time interval over specified conditions

• The ability for something to fail in a good way, without catastrophic consequences

For good reliability, engineers should focus on statistics, probability theory and reliability theories. Design for Reliability (DFR) is a discipline emerging that refers to the process of making the products or systems more reliable within a company.

Typically, the first step in Design for Reliability is to establish the reliability requirements that the system or product needs. During the design of the system, the requirements of great importance are placed in subsystems by the design and reliability engineers who work collaboratively. Once the product model is available, predictions are made about the lifetime of the components based on historical information. Many times, these predictions are not very precise. However, they are very useful to evaluate different alternatives of parts or models.

One way to increase reliability is to use redundancy, which will make the system less likely to date when using an emergency system that will keep it active. However, redundancy is often difficult and expensive and is therefore limited to critical parts of the system. Another technique is to study the physical faults in detail, either by mathematical models or through specialized software as a finite element to be able to predict the behavior of the material and to be able to redesign the part or product if necessary.

Many techniques can be used for reliability design, among which the following stand out:

- Built-in test (BIT)

- Failure mode and effects analysis (FMEA)

- Simulation of reliability modeling

- Thermal analysis

- Analysis of reliability block diagrams

- Analysis of Tree failure

- Root cause analysis

- Circuit analysis

- Accelerated tests

- Weibull analysis

- Electromagnetic analysis

- Statistical interference

The design for reliability is of great importance in Concurrent Engineering since it is very important to take into account the specifications that the product must meet to make certain changes when making its design, in such a way that it complies with the established requirements.

Development of the General Model

The proposed model integrates the philosophies of Lean Manufacturing, Design for Six Sigma, Six Sigma and Concurrent Engineering to carry out the routes of action, thus developing a new process, product or service, waste reduction, continuous improvement, inspection, resolution of problems; in such a way that the company can have a methodology that allows the development of a fast and efficient action plan.

According to the needs that the company has, the different possible routes should be followed. For the implementation of a new process or product, you must choose the Concurrent Engineering and Design for Six Sigma; if you want to increase production, correct defects, increase the quality of the product or service, reduce time or make continuous improvement, you should choose Lean Six Sigma; if you only need a solution to a problem or improve quality, you can simply choose Six Sigma; or in case of continuous improvement, reduction of waste and time, you can choose the Lean Manufacturing.

Development of New Process or Product

Before entering fully into the Concurrent Engineering and Design for Six Sigma, the following approaches should be made:

When entering a new process, the company should consider if what you want is to keep a current product or if what you need is to create a new product. In both cases, the company must evaluate through market studies if what it wants is to conserve the market that it currently has or if it wants to expand to new markets. In the case of creating a new product in the current market, you can do it thinking of new business opportunities to increase sales; you should also consider the possibility of expanding to a new market through innovation by creating a new product that can be used by more people.

The company may also choose to keep the product outline, making small adjustments to make it more competitive in the current market, for this adjustments or modifications to the existing process must be made. The other option of the company will be to reach new markets through these same adjustments to its product or through new alternatives for its export. Once you have very well established what you want, you can continue on the path of Concurrent Engineering and Design for Six Sigma, either to create a new product totally different from the previous ones or to modify the current process or product.

As described above, in the case of wanting to make continuous improvement of an existing process and quality improvement, the most convenient is the integration of Lean Manufacturing with Six Sigma. Once implementing any of these Methodologies and Philosophies, the results obtained should be followed very closely, in order to be able to tell if the expectations of the project have been met or if it is necessary to redesign completely the process. In case that the expectations are being fulfilled and there is no need to redesign the process, we must apply the relevant controls so that our production is maintained at a good level.

Customer Relations / Suppliers

The relationship with the suppliers is of utmost importance in all the stages of Design for Six Sigma and Lean Six Sigma, it must have an excellent relationship with suppliers if it is desired to expand the company. It is even more important when you want to become a leader in Lean Manufacturing, as the raw material must arrive in the required quantities, on the date needed and in the form in which it is requested.

The company should communicate regularly and clearly with suppliers; it should not be assumed that the manufacturer or supplier understood what we need. Clear expectations must be left in writing, without any ambiguity, especially in terms of quality so that the company can follow up.

In case of implementing Pull (Jalar) type systems, the company must let its suppliers know that they are working with this system and invite them to carry out the same type of system within their organization in order to reduce all the times of Wait associated with operations of which our company has no control. We must work in communication and negotiations with suppliers so that we can have the amount of material we need when we request it, especially when dealing with small quantities in order to reduce the excesses of inventories. The company must perform periodic evaluations of the suppliers in order to ensure that they are able to meet the expectations of the company.

Implementation of Knowledge Database

It is very important to carry proper documentation in all phases of the project to be able to use it later when carrying out new projects or improvements. That is why it is essential, sooner or later, the creation of a database in which all these documents can be saved, which allow us to carry out analyses of future projects and to have a better prediction of the costs that may arise in new projects.

The biggest advantages of using a knowledge database are the following:

• The staff is able to find and send information in a faster way.

• Decision making quickly.

• Reduce employee training time

• Retention of intellectual property.

• Departments and employees work more efficiently.

We must remember that the accumulation of knowledge is key to the success of the company.

Design for Six Sigma (DFSS) / Concurrent Engineering (CE)

The model is based on the ICOV method (Identify, Concept, Optimize and Validate), adding a Road Map according to certain conditions that must be met in order to proceed with the following stages, in addition to adding Concurrent Engineering tools which will help the development of more efficient products, with fewer defects and with lower production costs.

In general, it is about complementing all the benefits of Design for Six Sigma with the advantages offered by the tools of Concurrent Engineering for the creation of new products and processes.

Stage 1: Creation of the idea.

In this stage, we realize why the project was conceived in the beginning, what are the goals of the project and what are the resources that are available.

The following steps should be developed:

• Identify the expected market or potential customers

• Advice to take advantage of different marketplaces

• An estimated development cost

- Risk assessment

- Verify the necessary funds available to define the needs of the client

- Identification of leaders who can carry out the project

- Review bibliography

Stage 2: Voice of the client and the company

The voice of the customer (VOC) is very important when developing new products since this information tells us in an appropriate way the way to take based on the needs of the client and the company.

When you reach this stage, you must already have the approval of the people who will supervise the team that will carry out the project. The steps to follow are these:

- Creation of an essay that includes project objectives, metrics, resources, design parameters, and team members.

- Complete a market survey to determine the customer's needs that are critical for the quality and satisfaction of the client (Voice of the VOC client). When the client's needs are collected, they must be analyzed with the help of QFD and Kano analysis. Once analyzed, the most appropriate metrics (CTQ) should be chosen in order to measure and evaluate the design.

- Establish minimum requirements.

- Translate the Voice of the client into the critical metrics for quality, cost, and shipping.

- Establish "acceptable" performance levels for the new design.

- Advice on the required technologies.

• Project development plan.

• Align the project with the objectives of the company.

• Establish a reasonable price for the product.

Stage 3: Concept Development

To reach this stage, you must already have the approval of the project supervisors, in this part, you must estimate the target cost you want to reach, in order to determine the target cost, you must proceed before:

• Define the technical and operational requirements of the system. We can use the axiomatic design to transform the client's wishes into requirements.

• Obtain the conceptual design of the process or service.

• Generate different alternatives (trade-off). We must develop the parts or functions that the product must have so that it can work with the previously established requirements. It must be evaluated if our current technology is capable of satisfying the requirements; if it is not capable, a new process design must be developed. We can use the Axiomatic Design and the TRIZ for this case.

• Evaluate the alternatives (trade-off). Once we have different alternatives, we need to evaluate them, for this, we can use Pugh's selection techniques, Design Review, and Failure Mode Analysis to determine the weaknesses of the alternatives.

• A plan for the management of the project, with a timetable for the project where the necessary tests are included.

• Expected earnings and market growth.

• Review of suppliers.

• Verify that adequate funds are available to proceed with the preliminary design.

Many concurrent engineering tools can be used in this stage such as the design for Manufacturing and Assembly (DFM / A), Design for Service, Design for Reliability, product simplification, packaging optimization among others.

Stage 4: Preliminary design

The following should be done at this stage:

• Development of the objectives of the tests under nominal conditions

• Design, performance and appropriate transfer functions

• Report design analysis

• Risk assessment

• Approval of the selected design

• Action plan to follow with the selected design

In this stage, a validation can be done through software simulations of the model that will have to be verified. Otherwise, the parameters of the simulation must be reviewed to be carried out again. If the current estimated cost of the model is less than the target cost that we originally proposed and if the model is valid and goes according to what we need, we continue with the next stage, otherwise we return to the concept stage and correct it.

Stage 5: Optimization

Once the preliminary design has been approved, its performance, cost, and handling of parts must be evaluated in detail. In design optimization we have the following:

• Define all design documentation (Parties, operations, technology providers, costs, etc.).

• Make sure that the performance and functionality of the product or service meet the customer's requirements under the different operating conditions.

• Optimize transfer functions. For this, we can use the Design of Experiments (DOE) to quantify the transfer functions between the requirements and the critical factors. We see how the values of the outputs change with different inputs, and we can choose the best relation according to the requirements.

• Design of parameters and tolerances.

In this stage, we remove all the parts that do not add value to the product and try to use standard components to simplify the design. We use tools such as Failure Mode Analysis, Weibull Analysis, Thermal Analysis, among others.

Stage 6: Validate the design

For this phase, we must already have the drawings of the parts of the product, tolerances, process plan, feasibility, costs, suppliers and control plans. Having all this proceed to validate the design:

• With pilot tests that indicate the capacity of the process and where we can do the different statistical techniques to evaluate times, waste, amount of raw material needed, etc. With this, we can evaluate more accurately the performance in real life.

• Verification with a computational model.

• Validation of the control process

Stage 7: Previous arrangements, Production, and Documentation

The necessary adjustments must be made to obtain the greatest possible efficiency, and in case there is a problem, look for a solution quickly. Already in this last stage must have the necessary equipment to start production (people, raw materials, machinery, technology and infrastructure), all documentation must be attached to a database (Knowledge Base) for different purposes in the future, this Information should include all the project information, from the equipment, calendars, materials, suppliers, different designs, analysis results, problems found, costs, etc. The relevant statistical controls and the necessary tests must be put into operation (confidence analysis, error proof, process capacity, etc.).

Lean / Six Sigma Manufacturing (Lean / Six Sigma)

When we want to modify an existing process, either to solve a problem or for continuous improvement, we select Lean / Six Sigma (Lean Manufacturing / Six Sigma). The following process is a combination of the Six Sigma DMAIC methodology and the Lean Manufacturing techniques, combining both the focus on quality (reducing variability and increasing the sigma level) and the added value (reducing waste). and increasing efficiency). It must be taken into account that in this system, the critical quality requirements (CTQs) must be defined in Lean Manufacturing measures (waiting times, WIP, productivity, etc.). The combination of the Lean philosophy with the Six Sigma methodology allows an integration that will help the optimization of the entire process, since the Lean Manufacturing alone cannot take a process under statistical control, in addition, with both costs are reduced and the speed of the process is improved.

Doing the integration of this model, we begin by:

DEFINE

Steps to define in Lean-Six Sigma

The definition stage includes several tools and methods such as those shown in Figure 3.3, the reports of this stage must include the expected scope and objectives of the project, benefits, ownership of the project, structure, cost and critical quality requirements (CTQs). This stage includes the collection of lean measures to have more information about the process so that the team in charge of the project can start with a clearer vision.

The waste that exists must be understood so that the corresponding parameters are included within the process variables that affect the time and yield of the process. The project that is chosen must be a project that satisfies the critical factors for the clients and for the business. Brainstorming, Balanced ScoreCard, customer feedback or Value Chain Mapping can be used to choose the project.

The members of the team must be selected wisely; they must be familiar with the methodologies or philosophies to be implemented and must be able to influence others and know how to perform good teamwork.

TO SIZE

The measurement stage also involves several tools and methods such as those shown in Figure 3.4. The deliverables in this stage should include the Mapping of the Current Value chain (VSM) with the Lean measures (elements, process, flow, and information) as well as a simulation of discrete events (DES), which includes the logic of the process, animation and current performance. This will allow the process to be represented as close as possible to real life. It is very important at this stage to identify the sources of variation and waste.

A plan for data collection should be developed, and the

stability of the process verified to later calculate the capacity of the same. In this stage all the opportunities for improvement that come from the Map of the Value Chain are identified; the problem is understood to later identify the tools that help us improve the process.

ANALYZE

Once having the necessary measures proceed to the analysis. The methods and tools used in this stage are those shown in Figure 3.5. The reports in this stage should include a detailed understanding of the critical parameters of the process and the flow elements involved.

The Design of Experiments and regression models are important in this phase to see at what level it is appropriate to put the parameters of the process so that they meet the critical parameters of quality with the least possible variation.

In this stage, the process flow, the lead time and the production rate (Takt Time) are analyzed. The causes of the Process Map are identified and then grouped with an affinity diagram. With tools such as Matrix Cause-Effect the main causes are reduced to those that impact more on the problem.

Once the causes have been identified, the improvement tools are chosen, in such a way that Lean tools will be used mostly if the biggest causes of the problem are the times; or Six Sigma tools if the biggest causes of the problem are the defects.

TO GET BETTER

In the improvement stage, a better design of the process should be proposed. The methods and tools are those used in figure 3.6, and the deliverables should include: a mapping of the value chain of the future state with measures of proven improvements and an action plan. The actions and measures include the changes made as a result of the optimization of the Design for Experiments (DOE) and the implementation of the slender techniques. This is a stage where several lean techniques are implemented to restructure the process and obtain a better performance.

It should focus on the elimination of actions and elements that do not add value to the system, reducing waiting time and excessive inventory. Improvement solutions are generated and evaluated to later optimize them and be able to carry out pilot tests.

In this stage, all the data that the analysis phase gave us is used to generate solutions to the performance of the process; an implementation plan must be developed to monitor and follow the tools according to the times that we set.

CONTROL

Once the improvements implemented to meet the objectives presented above, the control stage should be followed. Otherwise they should look for other lean techniques that satisfy them or completely redesign the process if required, seeking a solution Design for Six Sigma (DFSS) / Concurrent Engineering (CE).

In this stage, once again Lean and SS methods and tools will be used, as shown in figure 3.6. Deliverables at this point should include: A Control and Monitoring plan that is aimed at improving the level of performance, performance and making the process robust and flexible for future changes. This seeks to maintain the performance of the Lean-Six Sigma integration. Sensitivity analysis using a simulator help monitor and develop plans and measures that are effective for sustainable improvement.

You can also carry out Mappings in the Value Chain in order to measure the achievements that were obtained with the implementation of the improvement and control tools and be able to compare them with those we had previously foreseen.

The Control Plan must make clear all the critical variables of the process that were identified in the previous stages in order to propose methods to control and monitor them in the future as the needs of the company change.

BIBLIOGRAPHY

What is Six Sigma? (2019). Lean Manufacturing and Six Sigma Definitions. Retrieved 10 May 2019, from http://lean-sixsigmadefinition.com/glossary/six-sigma/

Six Sigma Certifications - Starting Only $49 - Affordable Certifications, Free Books! 50% Off Until Early Next Week! (2019). International Six Sigma Institute. Retrieved 10 May 2019, from https://www.sixsigma-institute.org/What_Is_Six_Sigma.php

7 Reasons to Get a Six Sigma Certification. (2015). Simplilearn.com. Retrieved 10 May 2019, from https://www.simplilearn.com/reasons-to-do-six-sigma-certification-article

Six Sigma Definition - What is Lean Six Sigma? | ASQ. (2019). Asq.org. Retrieved 10 May 2019, from https://asq.org/quality-resources/six-sigma

A Brief Introduction to Lean, Six Sigma and Lean Six Sigma. (2019). GreyCampus. Retrieved 10 May 2019, from https://www.greycampus.com/blog/quality-management/a-brief-introduction-to-lean-and-six-sigma-and-lean-six-sigma

Larry Goldman, M. (2019). New to Lean Six Sigma? Moresteam.com. Retrieved 10 May 2019, from https://www.moresteam.com/new-to-lean-six-sigma.cfm

Six Sigma: overview, definitions and techniques - Business-Balls.com. (2019). Businessballs.com. Retrieved 10 May 2019, from https://www.businessballs.com/performance-management/six-sigma-definitions-history-overview/

The Importance of Six Sigma Performance Measurement. (2010). Bright Hub PM. Retrieved 10 May 2019, from https://www.brighthubpm.com/six-sigma/70094-six-sigma-performance-measurement/

LEAN SIX SIGMA

THE ULTIMATE ADVANCED GUIDE TO LEARN & MASTER LEAN SIX SIGMA

INTRODUCTION

By now, you have the basics of Lean Six Sigma. In fact, you might have a Green Belt and are preparing for your first job as a Lean Six Sigma professional. This is an exciting, yet nerve-wracking time as you have so many questions and worries going through your mind. Like most people who are starting their first professional job, they wonder what will happen next. They worry about how they can "prove" themselves as a valuable Lean Six Sigma professional. You might have a few Lean Six Sigma jobs under your belt but want to make sure that you are doing everything you can to make the process run smoothly for the company, customers, and yourself.

Lean Six Sigma is a growing profession that is showing up in various companies. Some companies are making Lean Six Sigma part of their team permanently, while other companies are requesting professionals to come in temporarily to help with certain projects. No matter what type of job you take on as a Lean Six Sigma professional, there are a lot of details involved. You need to work alongside a lot of people who understand Lean Six Sigma and some who do not understand. You will work with customers, stakeholders, your Lean Six Sigma seniors, and many other people. All of these people will help make the projects successful, as long as everyone is working together and doing their best. Even after a project is completed, there is still the continuous data collection to ensure everything continues to run smoothly.

Chapter one starts with some of the basics you didn't learn about in the previous books. You'll learn about a few career opportunities for a Lean Six Sigma professional. For example, you will learn the typical job duties for a senior project manager, a Lean Six Sigma consultant, director of operational excellence, business process manager, and Six Sigma analyst. It is important to keep in mind there are several career opportunities for you and this book only discusses some of the most common. You will also receive a few advanced tips that will help you thrive from day one on your new job and common issues with implementing Lean Six Sigma.

Chapter two takes a deeper look at data collection and statistics. As you know, there are five main steps to the Lean Six Sigma process. While they are all extremely important, data collection is one of the most challenging. This chapter will discuss several types of data and focus more on the planning that you will conduct on your Lean Six Sigma team. You need to be careful when you focus on the planning of your project as it will set the tone and the pace for the rest of your project.

Chapter three takes a look at analysis. Like collecting data, this is another important step of the Lean Six Sigma process and is also one of the most detailed. Many people become frustrated at this stage because they find themselves going back to collecting more data. It is important to realize this is going to happen. Even some of the most experienced Lean

Six Sigma professionals miss some data collection or don't have enough when they start analyzing the collected data. The best step to take is to go back and collect the data you need to give the company the best possible outcome. Not only will you learn why data analysis is important and what you should look at when you are on the job, but you will also learn about various types of analysis you will use. Finally, you will receive the five steps that will give you the clearest data analysis possible. You always want to keep these steps in mind as you are analyzing your data. Even if you feel you already know the outcome, continue to analyze your data clearly.

Chapter four brings you into the first case study. Healthcare is a rapidly growing field in the Lean Six Sigma world and one of the main reasons people are learning what Lean Six Sigma can do for their business. This case study focuses on a medical center that wanted to improve their kidney transplant area. Through this chapter, you will learn about the process the Lean Six Sigma team and the medical center followed in deciding what to measure, how to analyze, what they will do for improvement, and how they continue to control their outcome.

Chapter five is another case study that focuses on recycling and waste management. First you will learn about why recycling is beneficial and how it fits in with Lean Six Sigma. You will learn about two successful Lean Six Sigma case studies. The first study focuses more on waste management at a college. The second study focuses more on implying Lean Six Sigma practices in a government recycling facility. Each study will explain the process they went through and how they continue to control the practices implemented.

Chapter six is another case study that focuses on education. At first, most people didn't think that Lean Six Sigma practices could help education, but this is quickly proving to be wrong. In fact, Lean Six Sigma practices can improve education in many ways—for the administration, students, and the community. This specific case study focuses on a midwestern university that suffered from low enrollment and poor student satisfaction. Student services, student focus groups, and the business department all agreed that the best step was to bring in Lean Six Sigma professionals to help the university implement their practices. Throughout the Lean Six Sigma process, students quickly started to notice their college experience improving. You will walk through the process and learn about the changes made for this college.

Chapter seven focuses on using Lean Six Sigma practices to develop a startup business. It is more common for business owners to conduct Lean Six Sigma practices in a business that is well established. However, many people state it is better to bring Lean Six Sigma practices into the business from the beginning. This chapter shows the benefits of taking this step and how a fictional business incorporated Lean Six Sigma practices into their business.

Chapter eight discusses the importance of continuing to manage Lean Six Sigma practices. As a Lean Six Sigma professional, you will often move on from the project at this point. However, you still remain a silent partner as the project team will contact you if there are any future problems or if they have a concern. You also want to take charge of your projects and make sure that everything is running smoothly after the project is complete. For example, you might contact the business a year after the completion of the project to see where the business stands at that moment and compare it to where it was when you started. This can help you improve your own methods and grow as you continue to help other businesses become just as successful. Some of the methods you'll learn in this chapter will focus on how to handle conflict. Conflict in the workplace is something that can cause any planning to be put on hold as people don't want to work together when they are in conflict with each other. It will also break down communication and can quickly cause more problems. Therefore, you need to efficiently and quickly take control of conflict through a series of important steps.

If you have any questions or wonder what the process for Lean Six Sigma professionals is like, this book is the right one for you. You will go beyond the basics of Lean Six Sigma and look into what a work day is like for a Lean Six Sigma professional. This book will help you stop worrying about your next Lean Six Sigma job and start focusing on how you can help the company reach their goals. You will gain more confidence in your field through the case studies and believe that you can succeed as a Lean Six Sigma professional, even if you decide to start your own business.

1

IMPLEMENTING SIX SIGMA IN THE WORKING WORLD

You have already learned the basics of Six Sigma training. You understand the levels of training and how each has its own benefits. Now, with your training, it is time to finally dive into the working world.

Like many other people who recently finished school or a type of training, you may be wondering, "what next?" While you may have a career goal in mind, such as healthcare, it is important that you understand all of your options. This will help you decide which career choice is the best for you, your level of training, and your skills.

Career Opportunities

There are many careers that benefit from Six Sigma training. In fact, many businesses are looking to hire people who have a background in Six Sigma as full-time employees. Businesses are moving beyond the temporary hires as they feel the skills and knowledge people learn from Six Sigma can carry the business further. They can help other employees understand the importance of correct spending, policies, and procedures. The main characteristics of a Six Sigma professional include lean management, zero waste, high quality, and continuous improvement. They are data-driven and produce tools and methods that are nearly perfect for the processes of a certain company.

It is important to note that while you have the skills, it is up to the business to establish the requirements for a Six Sigma position. This can be challenging for many people as they don't have the training or understand what a person with the proper training is fully capable of. Furthermore, the topic can be confusing when it comes to the various levels. Because of this, there are many businesses who will put new positions on a probationary period. This isn't because they don't think you won't do well as part of their company. After all, they hired you, so they believe you will. It means that they want to establish a time frame to work with you and then re-evaluate the requirements for the position. During the re-evaluation process, you might bring information to them that they can tell you to utilize in your job. You will give them a better sense of direction as to how you can help their business thrive.

Probationary periods also depend on the company you work for. For example, universities and larger corporations will have probationary periods. But, smaller businesses that run on a small staff may work more closely with you from the beginning. Once you are hired, they might talk about your job, skills, and how everything will come together. You might work on your job description with your supervisor.

No matter what happens in the course of your career, you always need to keep in mind the skills you learned and do your best to help the company thrive. To help you get a better understanding of the job opportunities available, here are a few popular career paths.

Six Sigma Consultant

Many Six Sigma consultants start their own business. They will receive training and might look into working with a company that hires consultants to give them some experience, but many people have the dream of starting their own business. They will travel from one business to the next, sometimes remaining in a certain area and sometimes traveling all over the country, to help businesses take control of their policies and procedures. They will help conduct surveys so that the businesses have an idea of where to go from the start.

One of the best steps to take before you start your business is to have a business plan and understand how much to charge. Many consultants undercharge at the beginning because they are trying to establish themselves and believe they will gain more clients this way. Some will undercharge because they want to work specifically with nonprofit organizations and understand they can't afford to pay what corporations or for-profit operations can pay. However, you want to watch how much you limit yourself, especially in the beginning. Limiting yourself too much can cause you to lose valuable experience and exposure. Some consultants will set a certain price for nonprofits and charge a higher rate for corporations and for-profits.

Some of the requirements that companies want to see in their Six Sigma consultants include:

- Provide exceptional customer service skills
- Have extensive knowledge of Lean Six Sigma
- Have a problem-solving ability and understand the tools needed
- Serve as an advocate to ensure all business processes are running smoothly
- Coordinate all aspects of improving procedures
- Stay up-to-date on Six Sigma training
- Research and analyze organizational data so the company can provide the best working relationship with their clients

Six Sigma Analyst

Six Sigma Analysts are often referred to as business analysts, but they are different. Six sigma analysts receive different training. However, they focus on the same type of job duties as a business analyst. There are many people who start off their career as a business analyst and then turn to Six Sigma training to become a stronger analyst for their clients. Many business analysts won't work their way toward the Master Black Belt. They tend to receive the Green Belt certification. With this type of certification, business analysts will learn about run charts, statistical process control (SPC), box charts, Failure Modes and Effect Analysis (FMEA), and Measurement System Analysis (MSA).

Senior Project Manager

Senior project managers are either Black Belt or Master Black Belts. They support the company in many ways, such as launching brands, changing their brand, and strategies.

They help companies establish new policies and procedures. Some will spend their time helping their clients adjust to their new procedures while other senior project managers will request assistance from Green Belts for this task and make this their primary duty.

Senior project managers understand the importance of stakeholders and will help their clients strengthen their relationships with stakeholders. For example, they will set up meetings where their clients organize their stakeholders into specific categories, so they know who to contact for support on what project. They will learn techniques to keep their stakeholders interested in the company.

Senior project managers will manage more than one project at the time. This is different from Green Belts, who tend to take on one project at a time. The projects might be for the same company or they might travel from company to company. For example, senior project managers might spend two days out of the week with one company and the rest of the days with another company.

Senior project managers help to strengthen communication in the work environment and with the company's customers. They understand the importance of analyzing the communication through past experiences, research, and collecting data. They work closely with the leaders of the company to ensure that the processes are running smoothly and adjust where necessary.

Business Process Manager

Business process managers design, execute, monitor, evaluate, design, control, and measure business processes. They analyze the mission and goals of the company and compare them to the company's procedures and policies. If they don't match up, the business process manager will work with the company to create harmony between the mission, goals, procedures, and policies.

Many business process managers will start working with a company at the beginning. They will help them develop their procedures and policies and see everything through to the end. Because of this, they can spend months to years at one job. Sometimes they are hired full-time for an indefinite period of time by a company.

Business project managers have an eye for the bigger picture, but they never forget about all the pieces of the puzzle that create the larger picture. They help the company to understand that continuous change is important for growth. Business process managers believe that once a company stops changing and growing, they start losing sight of what is important and will find themselves struggling.

Business project managers who come into an established business will look closely at the flow of the company and rearrange any steps necessary to create a strong foundation. Business project managers are self-motivated to make sure the company they are working for succeeds in the long haul.

Six Sigma Project Manager

A Six Sigma project manager is similar to the business process manager, but the project manager focuses more on the projects of a company than its policies and procedures. The job outlook for a Six Sigma project manager is full of growth with about 87.7 million people working in this type of role by 2027 (Schembri, 2012). Keep in mind, this doesn't mean that you should aim to become a project manager after receiving your certification. You want to make certain that you will enjoy the position you choose. Therefore, it is more important to pay attention to a job description than growth and salary.

A Six Sigma project manager's main duty is to work on eliminating waste and problems within the company's projects. Other job responsibilities include:

- Planning a project
- Managing resource budgets
- Providing support for the company's employees
- Making sure everyone follows the timeline for the project
- Report regularly to stakeholders
- Problem-solve throughout the project
- Evaluate the outcome once the project is complete

Director of Operational Excellence

Director of Operational Excellence is another growing position that many companies want to bring into their team as they will help manage the stakeholders and continue to improve the company internally. Some of their most common job responsibilities include:

- Identify where the company can save within a certain project
- Create strategic plans that align with the company's policies and procedures
- Ensure that all policies and procedures follow any local, state, and federal laws
- Provide leadership to the company's employees when executing Six Sigma strategies
- Develop and track the benefits that Lean Six Sigma is bringing to the company
- Analyze outcomes and help the company maintain the Six Sigma strategies

Opening Your Own Practice

It is possible to set up your own practice with Lean Six Sigma. It doesn't matter if you have a Green Belt, Black Belt, or you are a Master Black Belt, as long as you have a clear business plan and a Six Sigma support team on your side, you can start to reach out to local businesses and offer your services. Before you take this leap, it is important that you have some experience working with businesses. You can use your experience to help sell yourself and show other businesses why they should hire you as their Lean Six Sigma professional. You also want to have a clear plan of action before stepping foot into anyone's business when you are on your own. What are your policies and procedures for your business? How many businesses will you focus on at once? Will you work with the business full-time or become more of a silent partner who makes meetings, supports, and encourages the company throughout the process? You want to think about all these factors and find a mentor with Lean Six Sigma experience to help you through this process.

Common Issues with Implementing

Most companies who have implemented Six Sigma professionals understand its benefits and will continue to work with Six Sigma professionals. But, this doesn't mean that the process started out smoothly. There is a lot that can go wrong when it comes to implementing Six Sigma in a company. Some of the factors fall on the company while others will fall on the Six Sigma employee. However, everyone on the team can overcome these challenges as long as they keep the lines of communication open, have mutual respect, and remain on the same page.

It will happen—you will make a mistake in your job. It doesn't matter if you are a new Six Sigma professional or you have been a part of the field for years. Mistakes happen on both sides and you have to know how to handle them professionally and personally. When it comes to mistakes, you want to ensure that you do what you can to fix the situation, learn from your mistakes, and move on. It's important to remember that you are in charge of you. You can't control what other team members do, how they react, or if they continuously remind you of the mistake. How you handle the mistake and the outcome shows people your professionalism.

Below are some of the most common challenges and mistakes that Six Sigma employees need to overcome during their process. The more mindful you are of what you are doing, the less likely you are to make a mistake. These are here to help you remain mindful of what is happening in your position and to work with the company to fix any mistakes before they become too much for some people to handle.

Determine the Customer

The customer isn't always clear when it comes to Lean Six Sigma implementation. In the beginning, any Six Sigma professionals make the mistake of thinking the customer is an employee of the company. While this can be the case if you are there to help a company with their policies and procedures, the customer is typically the company's customers or patients. For example, if you are helping a hospital's physicians see more patients every year, your customer is not the physician—they are the patient. You are working with the physicians or employees of a company and, in many cases, they are your colleagues as you are also employed with the company. Some Lean Six Sigma professionals look at the employees of the company as their clients and separate them from the customers.

Measuring the Impact

Measuring the impact of your marketing and how the changes are affecting the company, its policies, procedures, and customers is a challenge. One reason for this is because there is no common way to do so. There are many ways to measure the effects of your work, but there are also many areas that you need to look into. Each of these areas might have a different formula or report for you to analyze. Then, you need to focus on combining all the reports into one so you can get an overall idea.

Measuring the impact of your outcome can take a lot of time and patience. You will spend years looking at the results. For instance, if you spent your time helping an emergency room speed up its wait time, you might analyze the outcome over a period of three to five years. Throughout this time, you might look at areas the hospital can further improve in, which means you will need to go through all of your steps again. This brings more analysis that you will spend time on.

Relationships

Once you get your job, relationships will become your number one priority. You will need to secure relationships within the company where you work and with their customers, stakeholders, and anyone else associated with the company. Relationships will help you with your marketing and sales campaigns. It's important to remember that when you are trying to sell a product, it's not about the process you use—it is all about your relationship with the customer. If you have a strong relationship and they respect you, they are more likely to notice what you are selling and make purchases. If you don't have a good relationship with them, they don't care too much about what you are selling. They might feel that listening to you is a waste of time.

When it comes to relationships, respect and honesty are key.

You need to gain the customer's trust for them to respect you. Once you have accomplished this, you will have an easier time communicating with them and an overall pleasant experience. Gaining trust and respect can take a while, so you will need to have patience and understand that it might take a few conversations before you can talk to them about the product you are selling.

Poor Execution

One of the biggest mistakes that can cause issues with implementing is poor execution. This commonly happens for a variety of reasons.

1. You recently received your certification and you are on your first job. It's hard to know exactly what you need to do without communicating with someone who has more experience. Keep your lines of communication open with your senior manager as they can help you make sure your plan is strong from the beginning.
2. You don't understand the policies and procedures of the company. When your plan doesn't align well with the company's policies and procedures, you will have poor execution. You have to align your plan with the organization's goals.
3. You don't put the right amount of effort into your plan or work. Many experienced Lean Six Sigma professionals find themselves in a slump where they feel like everything is the same and they become too comfortable with their position. When this happens, they will miss important steps in their plan.

Always do what you can to establish a strong plan from the beginning. This will help everyone make sure that everything that is happening aligns with the organization's goals, mission, policies, and procedures. If something doesn't align, then it is time to go back to the drawing board and come up with a new plan, or at least a part of one.

People Don't Understand Lean Six Sigma

While Lean Six Sigma as a profession is not new, many people don't understand what you do, especially customers. When you tell them that you are Lean Six Sigma expert, they will question you and what your role in the company is. They will want to know why you are contacting them and not the owner or other employees of the company, especially if they are a repeat customer and have a good relationship with the company. This can cause issues at the beginning when you are trying to build a relationship. It is important to remember to remain patient and let the customer set the stage for what comes next. By giving them the lead, they will feel like they are in control. This is a technique used by many sales people. The more the customer feels like they control the situation, the more open they become. They start to trust and respect you more easily, leaving the door open for a positive and strong relationship.

Lack in Communication

Every employee becomes focused on their own goals and how the company benefits them. When this happens, you start to create a lack of communication because people don't think about what is best as a whole and for the company. Another way that communication is lost within a company is when people focus on what they are doing but don't report back. They also don't update other team members about any tasks or outcomes. The big problem with the lack of communication is the stakeholders. They are some of the most important customers and when they feel that the company isn't on the same page or doesn't communicate as they should, they will start to lose interest in the company.

It is hard to keep everyone on the same page and open about their progress. Because this often falls on you, this lack of communication is challenging and frustrating. The best way to handle lack of communication is to have procedures in place that ensures your team members are communicating with everyone. You might establish weekly meetings where everyone can update each other, have a daily email system, or have everyone give monthly reports.

Advanced Tips

Once you get into the Six Sigma field and take on your first job, here are a few tips you will want to keep in mind.

Don't Forget to Listen

You are the expert. The company called for your assistance because you are qualified in Lean Six Sigma. However, this doesn't mean that you stop listening to your clients. It is always important to ensure your clients feel their needs and emotions are met. This is one of the biggest steps in the Six Sigma profession that is often overlooked.

Even though you are skilled in Lean Six Sigma training and you have the certification, this doesn't mean that you know everything about the company. However, what you do know is how to help the company tighten their procedures and policies so they can help their customers. For example, you can analyze policies and notice ways to help the hospital's emergency room decrease the wait times for their patients. But, you still need to listen to what the hospital needs along the way.

When you push what your clients are saying about what is important in their company, you are creating holes in their business. These holes can cause problems when it comes any new policies and procedures that you aren't aware of until they start happening. This will not only cause problems for the company, but issues between you and the company.

Don't Expect a Typical Day

Unfortunately, there is really no "typical day" in the office for a Six Sigma professional. You need to be aware of all the situations that will need your assistance. Your level of training will determine the types of jobs you can reach for and how to conduct your day. For instance, Master Black Belts are the highest level. This makes them the most experienced Six Sigma professionals. Master Black Belts are basically a link between the executive leadership of a company and the Six Sigma improvement program. They will ensure the company receives the support they need, including funds, to implement any programs, policies, or procedures needed for an efficiently run company. Master Black Belts often ask for Black Belts to help them with projects as they tend to work for larger organizations where assistance is needed.

A Black Belt focuses on the Six Sigma improvement process. They often work under Master Black Belts, whether directly or indirectly. For instance, Black Belts might work in a different office, but their supervisor is a Master Black Belt. When processes become a little more complex, Black Belts will turn to Master Black Belts for assistance. Black Belts are seen as mentors and trainers for the company.

Green Belt professionals tend to work part-time. They will spend between 30% to 50% of their time focused on projects and the rest of their time working in their main functional area. Black Belts and Master Black Belts will often ask Green Belts to assist them with data and analysis. They will also help them expand their Six Sigma knowledge.

No matter what level you are, there are primary duties that every Six Sigma professional will conduct throughout their day. These include:

- Overseeing process improvement
- Facilitating Kaizen and Rapid Improvement Events (RIEs)
- Understanding improvement processes
- Helping your clients adapt to the improvement processes
- Assess customer needs and establish a stronger connection between your client and their customer base
- Develop and present formal presentations to senior management, such as where you are in the process and why you are taking this direction

Don't Stop Learning

One of the worst mistakes Lean Six Sigma professionals make is to believe they know it all. You want to keep your eyes and ears open because you will learn on the job just as much as you learned while getting your certification. You will learn from other Six Sigma Green and Black Belts, from the companies you work with, and from your customers. Walk into your job with the belief that you will learn something new and you can improve your skills every day. This is one of the first steps that you will make as a successful Lean Six Sigma professional.

Always Have Support by Your Side

No matter where you go or how long you have been in the field, you want to have advocates at your side. Your job will get tough sometimes and it will help to have a support team to call and help you through difficult situations. You will find support in the people you receive your certification and training from, senior management, sponsors, executives, and project managers. Your supporters can help you get a start in the field, keep you on the right track, answer your questions, and give you general support.

2

DATA COLLECTION AND STATISTICS

As a Lean Six Sigma professional, you will spend a lot of time collecting and looking at data. When talking about data, it is important that you understand it is not just about the numbers. There are various forms of research such as surveys, reports, and anything else you will collect to help the company improve their methods.

You already understand the DMAIC process. Now, it is time to look at this process as you are stepping foot into the field.

Types of Data

There are various types of data that you will look at throughout your process. You might focus on one specific type of data or you might look at different types of data. You need to focus on what follows your plan and other factors within the company and your job.

Location

You always want to pay attention to your location. You will do this at the beginning of your job. In fact, you might find yourself analyzing and researching the location of the company before you start your position. Location is important because it will give you a little background into your customers, stakeholders, team members, and everyone else associated with the business.

Location is important because large areas house more factors than rural areas. For example, if you are helping a local company expand into the online world in a community of 5,000 people, you will have different factors to consider than a company who is located in a city with 100,000 people. First, the products that the company with the smaller population has might not be suitable for a larger market. They might not have policies or procedures in place to handle a large sale. This is typically not the case for companies that are used to hundreds of people going through their doors every day.

If the company is already online, you will need to look at statistics that show the most popular areas. For instance, if the company has a Facebook page, you can look at their business statistics as they will tell you where the majority of their customers are located, whether they are in the United States or another country.

You can think of the location as one of the foundations when it comes to data collection. It is a great place to start your research.

Discrete Data

Discrete data can only handle certain values. It is a type of quantitative data. It is more specific than other types of data. This type of data is placed in categories, it doesn't matter how many, and this is where the data remains. You won't find data jumping from one category to the next. Once it is in place, it is settled. For example, the number of dimples on a golf ball is discrete data. This number will not change. Any type of counted data is discrete, and the number can be infinite.

It is important to note that if you have discrete data as part of your analysis and other team members focus on discrete data, people will have different numbers. This is normal and you should not become concerned.

There are three main types of discrete data:

1. **Nominal data.** This is a type of descriptive data and doesn't have much to do with numbers. Nominal data has more than two categories, such as color, phone number, age, and location. Nominal data doesn't have a ranking system.
2. **Ordinal data.** Ordinal data is put in a certain order. If you have a ranking system with your discrete data, you will use this type of data. This type of data cannot be measured. It can only be placed in a ranking system, such as first, second, third, etc.
3. **Binary data.** This type of data has two categories, such as yes and no or pass and fail. It is qualitative, which means that it doesn't deal with numbers.

Qualitative Data

Qualitative data is not a number and is anything that you might analyze, such as color or shapes. You can observe and record this type of data. For example, you are sitting in the lobby of the company and you observe the cashiers talking to the customers. This is a form of qualitative data you are collecting. You might be looking at how the company can improve their customer service. If you meet with the company's stakeholders or send customers surveys, you are performing qualitative data because you will analyze the answers you receive.

Some people will skip over qualitative data because they don't feel it is necessary. Qualitative data is just as important as any other type of data. When you observe situations, actions, or tone of voice, you are finding out the traits or characteristics of the company. This will help you establish a foundation when it comes to knowing what is working and where the company can improve their methods.

Qualitative data is all about the emotions of people. You learn what they feel, and you can take this and contribute it to your research. You will learn how customers and employees speak. You will learn the day to day interactions that the company's employees don't think about because they are natural.

Quantitative Data

Quantitative data is the type of data that focuses on numbers and formulas. You can use quantitative data when you ask questions such as "How many people live in each neighborhood?" or "How much does the company make in sales every day?" You can further divide quantitative data down to look into daily sales per category.

Continuous Data

Another name for continuous data is variable data. This is another type of quantitative data because it focuses on numbers. It is often preferred over discrete data because you can do more with continuous data. Because it is data that is constantly giving you more information, such as daily sales, you can break the number down into different categories. For example, you can break the daily sales down into clothing, household items, and crafts. You can also break clothing down further into women's, men's, and children's clothing or by jeans, shoes, and tops.

When gathering your data, it is always important to keep the DMAIC process in mind. You always want to remember your formulas, measurement, and general analysis. There might be times where you create your own system or formula to help get a better understanding of your data. While this is rare, it will be something that you observe during the process. All of the data you collect will help you gain a better idea of how to help the company improve.

The data that you collect will always be a part of your planning phase. When it comes to Six Sigma professionals, you can find yourself developing different plans depending on the stage you are at within your job. For example, you might have a beginning plan that you establish within your first week on the job and share with everyone. Once you have collected some data, you might develop a different plan or add on to your previous plan.

Planning

When you start the planning phase, you can start to feel overwhelmed. There is a lot of detailed information that goes into planning, which can cause most people—no matter how much experience you have—to feel like your job is beyond your skills. When you come to this moment in your career, it is important to remember that you are growing as a Six Sigma consultant, which means you are learning as you gain experience. Feeling uncomfortable is not always a negative factor.

To help you work through your planning worries, here are a few factors to consider:

Bringing in Stakeholders

Stakeholders are an important part of the company and many will spend time helping the organization throughout their projects. In fact, stakeholders play an important role when looking at the success or failure of a project. For example, many Lean Six Sigma projects have failed because the team members did not identify the best stakeholders, ignored stakeholders' issues, did whatever the stakeholders wanted to do, and underestimated the power of stakeholders.

Before you start the project, you want to look at the relationship the company has with their stakeholders. If the relationship needs to become stronger, you want to focus on this part first. Stakeholder management is a very important part of the Lean Six Sigma business. You will spend hours communicating with stakeholders and ensuring that everyone is one the same page. If you are working with a smaller company who only has a few stakeholders, you might have to contact all of them to see if they want to become involved in the project. If you are working with a larger company that has hundreds to thousands of stakeholders, you will want to meet with the leaders who manage the stakeholders and discuss which stakeholders are best for the project.

When identifying and classifying stakeholders, you want to analyze each one. You can evaluate the stakeholder's previous commitments to the company and meet with them for a one-on-one interview. Ask yourself a series of questions, such as how interested they would be in this project or if they show any interest when you talk to them about the project. Once you have an idea of where stakeholders stand when it comes to the project, start identifying the best ones for it. You can also focus on what stakeholders would be best for other or future projects with the company. If you find that some stakeholders are losing interest in the company, you want to work on strengthening the relationship with them.

When you have the list of stakeholders you want to include in the project, you have to come up with a plan of communication and engagement. For instance, you will discuss how to ask the stakeholders to become a part of the process. You will also note how you will keep them engaged in the project. For example, how will you keep them informed about what is going on? Will you send them a weekly or monthly report? You always need to consider how much communication is too much when it comes to your stakeholders. You will have stakeholders that will want to attend every meeting and receive the reports on a weekly basis while other stakeholders will feel this is too much communication. They would rather receive communication once every month or every quarter.

You need to keep in mind that all stakeholders have to be informed of the project and remain informed. Therefore, if your project will take years, you can do this on an annual basis. If you have a project that will take a few months, look at communicating with all stakeholders on a monthly basis.

Start with Goals

Before you go too far into your planning, you want to establish some goals. You will find out what goals are the best for the company in several ways. For example, you will read through the company's policies and procedures, talk to the employees, talk to the customers, observe customer services, look at previous data, and read through your job description.

When you focus on goals, you don't just want to look at any goals. You want to focus on setting SMART goals. This means you will look at lifetime goals, set smaller goals, and monitor goals.

SMART goals mean you will focus on five points:

1. Significant
2. Meaningful
3. Action-oriented
4. Rewarding
5. Trackable

For instance, instead of having a goal that states, "Decrease the wait time for patients," you would write, "To decrease the wait time for patients in the emergency room while improving patient care."

Along with focusing on the five factors of SMART, you also want to establish smaller steps within your larger goal. For instance, you will discuss the steps you will use to decrease the wait time and also improve patient care. You might look at this as two separate goals before combining them into one. You will also ask yourself questions like, "How can we decrease wait times?" "Should we hire more physicians?" "Are we short staffed with nurses?" and "How is our check-in process?" All of these questions can help you get an understanding of factors that contribute to longer wait times in the emergency room.

What are the Needs?

Before you start building your plan for the project, you need to assess the needs of everyone involved. For example, if you are looking to decrease wait times for patients at a hospital's emergency room, you have to identify the needs for the hospital, physicians, nurses, and patients. You can start doing this by asking yourself a variety of questions, such as "What do patients think about current wait times?" "How efficient is the staff?" "Is the hospital understaffed?" "Is training strong?" Once you start looking into your first couple of questions, you will find yourself asking more questions, especially when you start analyzing any surveys or other forms of communication between yourself, patients, and hospital employees.

Establishing needs will help you with creating the necessary steps for your goals. There will be some goals where the steps to achieve them are obvious. People will quickly know what they have to do to reach these goals. There are other goals where the steps are not as obvious. You and your team members will question how to succeed in this goal. When this happens, you want to look at the needs of the company and its customers. By assessing the needs and knowing what goals need to be met, you will build the steps to achieve the goal.

Begin Brainstorming and Close the Gap

This is the step where most people, especially organizations, want to start. They want to have a meeting where everyone sits down to discuss the areas of improvement. However, taking the time to discuss goals and needs are important prior to this step. You don't want to reach this step until you know what areas need your focus. Then, you will want to start by brainstorming. It is important to write down any thoughts that people have. It doesn't matter if it might not work out, the key is to write down whatever the team thinks might or might not work and narrow your ideas from there.

Once you have done the brainstorming part, you will move on to prioritizing the ideas and projects. You will create a list of projects from most important to least important. The importance of each project is based on the organization's goals and everyone's needs. If you get stuck or feel that you can't find a way to prioritize a few projects, ask yourself a series of questions that will help you understand the importance of each project. Remember, they have to align with the values of the company.

Dividing the Projects

Dividing the projects among team members is not always easy. There are a lot of factors to consider when dividing the projects. For instance, you want to think of everyone's unique skills. This isn't always the skills they were hired for. Instead, it's special skills that each person brings to the company that isn't dependent on their job. These skills are part of a person's personality. For example, someone who is strong in persuading people might get the job of talking to stakeholders for the project.

You also want to give team members projects that peak their interest. If they are given something that they don't want to do, they won't put their best effort into the project. They will do what they have to, and this will become good enough. But, when someone enjoys what they are doing, they put as much effort into the project as possible.

When you are looking into dividing projects, you want to think about who will give you a steady stream of results. This will help you when it comes time to stabilizing the project's execution plan.

Project Execution

First, you will want to think about the methods you will use to collect and analyze data for the project. You might ask yourself what the best continuous improvement methods are when it comes to healthcare, recycling, or any other project. You want to focus on efficiency and choose the best method that fits the problem.

When you start establishing your timeline, you want to keep it as realistic as possible. For example, if you take the average time DMAIC projects take to deliver results, you want people to focus about 200 hours total on the project every week. This means if you have 20 team members, everyone needs to work on the project for 10 hours every week. This doesn't mean that you give your team members double the work if you only have 10 members. It means that you will need to analyze what more they can do, if anything, from the 10 hours. You can also discuss how to bring more people onto the team. For example, would stakeholders or other staff members be interested in joining the team?

You need to make sure that everyone has the resources they need to make the project successful. For example, ensure that each team member can dedicate 10 hours to their week, that they have all the reports they need, they have any type of lists or funding necessary, and anything else that will help reach the goals.

Confirm that each team member understands how they will execute their part of the plan and how they will record any results they get throughout the week. Determine how each member will notify you or the leader of the project of their weekly results. For example, will every member write out a weekly report to show their results or will you receive copies of all the documents?

Part of your project's execution is to develop an outcome plan. One of the biggest mistakes beginners make is establishing a plan while everyone is working on the project but not garnering any results. One of the best ways that will help you validate the project's progress is through recording the results for a period of time after the project's end date.

3

ANALYSIS

One of the most important parts of Lean Six Sigma is the analysis that you will use throughout the project. It is common to use more than one type of analysis. You will also find yourself using some types during the process and then changing your methods after the completion of the project to control the continuous research.

The first point to realize when it comes to analysis is it's not all numbers and statistics. While you will spend a lot of time crunching numbers, you will also observe your research. You will do your best to read between the lines to give yourself the best outcome. Analysis with observation isn't the strongest method for everyone, including Lean Six Sigma professionals. Some people are better with numbers while others are better with observing written words and piecing all the documents together like a puzzle. Because of this, you always need to remember that you do not have to do everything on your own. You will always have a team to help you throughout the project and you can always request the aid of a senior Lean Six Sigma professional or a Master Black Belt.

A second point to realize is you will always use some type of analysis. There will not be a time before, during, or after the project that you are not focusing on analysis. This is one of the strongest tools for Lean Six Sigma and needs to be used as thoroughly as possible.

Why Data Analysis Is Important in Business

If you have never stepped foot into a Lean Six Sigma professional occupation, you will quickly notice how important analysis is when it comes to business. You spent a lot of time during your certification training on analysis and there is a strong reason for this. Without a clear understanding of the analysis phase in Lean Six Sigma, the project will not succeed. You can't reach your goals without analyzing the data, one-on-one interviews, among other things. There are several reasons why data analysis is important for each project.

You Get to Know the Target Customers

In the business world, there are the general customers, which includes everyone, and target customers, who hold the main characteristics of the general customers. When you are thinking about customers in business, you want to focus on the target customer to give yourself an idea of what the customer needs, what they will like, how to communicate with them, and how to keep them interested in the business. When a customer is interested in the business, they will continue to support the company. This will not only keep the company's doors open, but it will also help the company understand the best direction to take for future growth. A business cannot survive without a strong connection to their customers which is one reason why Lean Six Sigma is becoming increasingly important in the business world.

Cut Operational Costs

One of the biggest reasons businesses request Lean Six

Sigma professionals on their team is because they want to focus on the best way to cut costs. You will use various analyses to look into the best ways of cutting costs. For example, if you are trying to help an educational center cut costs so they can provide students with better services, you will use quantitative and qualitative analysis. Each main type of analysis has several subtypes that can extend your efforts and give the school the best outcome.

Improve Advertising

By understanding what customers want, you can focus on creating better advertisement that will pull them in. Better advertising will also pull other potential customers in that have similar characteristics to your target customer. Another benefit of improving advertising is it allows you to cut costs as you increase your sales. By eliminating the advertising that most people don't notice, you won't spend money on advertising that won't bring people into the business.

Helps You Problem Solve

Problems can cause more issues for a business than many people realize. There are some problems that can leave businesses stopping production for a period of time and closing their doors. Oftentimes, the symptoms of problems show up before anyone notices. This happens because people become used to the daily or weekly tasks of their business and they aren't mindful about what is going on in their environment. Other times, problems arise suddenly. Analysis will not only help you notice the problems, but continuous analysis can help you find the problems because they become bigger. Through analysis, you can problem-solve to establish a more effective business.

Analysis Stage

Analysis is a phase in the DMAIC that is filled with tools to help Lean Six Sigma professionals spot problems throughout the process. Analysis also helps people determine the root cause of the problems and how to achieve your goals by fixing the problems.

When you analyze, you do not want to take a bunch of guesses and follow the one that is closest to the cause of your problem. You want to take your time and ensure you use the right tools and come up with the best solutions for the problem. Some of the tools you can use during your analysis stage include:

Value Stream Analysis

Value stream analysis is usually considered as one of the key tools when it comes to Lean Six Sigma analysis. Value stream analysis shows the steps you take when you deliver the service. There are a few graphs and maps that you can use to determine the value stream analysis for your project, such as a process map.

Once you have the whole process mapped out, you can use value stream analysis to find the waste elements and focus on tightening your project so you can deliver the best results effectively. For example, you might find that the emergency department at the hospital needs to add in a database to keep all their staff on track about a patient's whereabouts and if they are ready for any tests or waiting for results.

There are five main steps when it comes to value stream analysis:

1. Specify the value, which is the needs of the customer or a specific price.
2. Identify the value stream.
3. Create the value stream flow, which places the steps you will use to reach your goal in an even pattern where the steps naturally flow together.
4. Use pull scheduling instead of push scheduling. Pull scheduling means you are establishing a reason for customers to come to your service. You are pulling them into the businesses with your services and benefits instead of pushing them in. Pulling helps ensure the customers will stay with the organization.
5. Aim for perfection. When looking at perfection, you don't want to end the process. You want to find a way to continuously monitor the process. This will allow you to always improve in areas and notice other areas that need improvement.

It is important that you look beyond the company you are working for when you analyze value stream analysis. For example, if you are looking at ways hospitals can save money, you will look at other hospitals to see what they incorporated to save money.

Cause and Effect Diagram

The cause and effect diagram is sometimes called the fishbone diagram because of its shape. With the cause and effect diagram, you will analyze the symptoms and cause of the problem. While the diagram doesn't give you the history, it gives you the factors you need to work on to come up with a solution to the problem. The problem is always on the right side of the diagram and the causes are on the left side. You will have an arrow pointing to the problem and branches from the arrow which represent the causes leading to the problem. Lean Six Sigma professionals like to use the cause and effect diagram because it is easily laid out in a way everyone can understand the causes for the problem. Once you have the cause and effect diagram in place, the team can start to focus on how to solve the problem by eliminating as many causes as possible.

Regression Analysis

Regression analysis is a way to focus on finding the root cause. Where the cause and effect diagram focuses on all the causes for the problem, regression analysis typically focuses on one—the biggest cause. At the same time, this does not mean you won't come across other causes for the issue. You will often use different variables of interest to examine the relationship, which leads you to learning about more than one cause. You don't need to focus on all of the causes you find through regression analysis at once. In fact, it is best that you focus on the main cause and then look at the other causes later. Sometimes the minor causes will be eliminated when you focus on the root cause.

To get a clear understanding of regression analysis, you need to comprehend the independent and dependent variable. The dependent variable is the main point you are trying to predict or understand. The independent variable is more than one fact that influences the dependent variable. You always need to recognize the dependent variable before the independent variable to focus on regression analysis. For example, if you want to know the average monthly sales for a department within a store, you will take all of the sales numbers over the last three years and place all of the information into a chart. You will then use formulas to give you the average sales. One of the nice factors about regression analysis is you can focus on the yearly, quarterly, monthly, or weekly sales. You can then further your analysis by looking at your top sale weeks or months and figure out why people are purchasing more items during certain times of the year over other times.

Another reason why Lean Six Sigma professionals use regression analysis is because they can see if the theory will fit their data. This measurement helps you understand if you are on the right track or if you need to adjust your methods to receive the best possible outcome.

Exploratory Data Analysis

In every business, there are relationships that employees don't consider. While these relationships continue to exist, people don't build on them because they are unaware that they are a part of this relationship. However, this can quickly become damaging to a business as the other person, such as the customer or another employee, might feel that they aren't getting the benefits they should from the relationship. Instead of talking to the other person about it, most people stop supporting the business. This can have a snowball effect as the person decides to tell other people about the poor service they received at the business. One way to make sure people acknowledge all of their relationships and do what they can to help them develop stronger relationships is through exploratory data analysis.

It is important to note that the connection exploratory data analysis can bring to the table doesn't need to include a relationship between people. It can also be a relationship between products, services, or other factors within the business.

One of the most common methods of exploratory data analysis is a Pareto chart. You can use other graphics, such as a stem and leaf plot, box plot, and histogram. No matter what graphics you decide to use for this analysis, the struggle you will have is using it to predict the future. Even though you can never truly tell what the future holds, it is important to use analysis to help you understand the possibilities of the future. This is something that exploratory data analysis will not do. It tends to focus on the past and the present. Therefore, to analyze the business' future possibilities, you need to look at other types of analysis.

One of the benefits of exploratory data analysis is you can get various software to help you conduct this process. Some of the software includes Python, Data Applied, and JMP. Another name for exploratory data analysis is causal data analysis.

Predictive Data Analysis

One type of analysis to include if you are going to use exploratory data analysis is predictive data analysis. As the name suggests, this type of analysis looks at historical and current data to give you a prediction for the future. The key here is you will be able to notice a trend within the data. This will allow you to drag the trend into the future. Furthermore, you will note what you need to do to ensure that your trend continues as much as possible. While the future trend might not look exactly the way you predicted it, they are typically close.

The downfall is that this is not a type of analysis that is easy and quick to understand. It takes a lot of technical expertise but provides great results. Many people will place their data into a graph as this allows them to see the trend easier. Through a line or bar graph, you can quickly pick out where the trend is at its height and where it is at its lowest. This allows you to focus on what leads to the trends highs and lows. For instance, if you notice that certain sales bring in more people, then you can focus on creating more sales like those. You may also notice that some sales don't generate a lot of customers. In this case, you will want to decrease or eliminate those types of sales.

Descriptive Analysis

Descriptive analysis describes the data. It can help people learn to understand the data they are observing and connect it all together. It is often the first step in the analysis because it involves finding mistakes, including typos, and helps you see how you can further analyze your data.

After you have a clear understanding of the data, you will decide what variables and approaches you will use for your analysis. For example, will you focus on quantitative or qualitative variables? Descriptive analysis can help you lay out your plan for the rest of your study.

Steps to Give You Clear Data Analysis

Data analysis is not easy to learn. Even if you have Lean Six Sigma training, you will learn a lot about data analysis while you are on the job. This is because data analysis is a hands-on position that requires a lot of expertise. It is hard to explain or teach people how to analyze data as it takes different types of thought processes and observations. It is something that you learn as you continue in your field. You will also find that you use your own techniques and strategies when you are analyzing data. For example, some people need a visual whereas other people can think about the data and come up with a solution. The way you analyze data doesn't matter. What matters is your understanding of the analysis method and how careful you are during the method. It is always better to take your time and be patient as you are focusing on analysis. When you perform a quick analysis, you can find yourself missing important information.

No matter how long you have been in the Lean Six Sigma field, you are constantly developing your analysis skills. Here are five of the best steps to make sure you are thorough and clear in your analysis process.

Step One: Always Make Sure You Define All of Your Questions

It is common for a Lean Six Sigma professional to come to the first meeting and receive a series of questions that the business wants to focus on. While this is a great start, these questions are usually unclear and not extremely descriptive. In fact, it is common for a Lean Six Sigma professional to question the questions as they don't understand what the business is asking.

You want to ask questions that will present or eliminate a solution. One way to do this is to start with the problem. For example, students at the local college are complaining they don't receive adequate advising. From there, you will focus the problem into a question by thinking about a solution. It doesn't matter if you feel the solution will solve the problem or not. The goal at this time is to develop clear questions. Therefore, an example of a question you might ask is, "Do the advisors have enough resources to provide the students with adequate advising?"

You can come up with as many questions as you are comfortable with. You don't want to limit yourself, especially in the beginning. If you find that you have limited yourself down the road, then you need to go back and reanalyze the step to formulate better questions.

Step Two: Make Sure Your Measurement Priorities Are Clear and Have Focus

Because there are different types of analysis, it is important that you focus on one and prioritize. To help you accomplish this task, there are two questions you can ask yourself:

1. What am I going to measure?
2. How will I measure it?

When you are deciding what you are going to measure, focus on your questions you created in step one. What will you need to know to answer your question? In the case of advisors having enough resources to adequately help their students, you have to know what resources the advisors have. You need to know the advising process and what the students think about the process. Throughout this process, you might ask yourself other questions that are related to your main question. If you ask yourself subquestions, it is important that you receive an answer for these questions as well. It is always a good idea to get more information than less information. You can always set information aside if you aren't going to use it at that moment. However, if you find you don't have enough information, you need to go back and collect some more, and this will take more time.

When you look at how you will measure the question, you will ask yourself another series of questions. For example, you might ask, "What is the time frame?" For instance, are you measuring time by week, quarter, or semester? You can also ask yourself, "What factors will I need to include?" To answer this question, think about the advisor's availability throughout their day and with students. You can also ask how much time the students have to talk to their advisor, including how many hours they are available.

Step Three: Collect the Data

Step three is to collect all the data you will need to answer your question. To help you stay on track with your data, it is important to focus on a few factors:

1. Use an interview template if you need to gather

information when meeting people for an interview. It is always a good idea to create this template before you begin the data collection process as it will help you save time. You can use this template to ask questions over the phone, in person, or through email. A template will also keep you and everyone else on the team consistent and ensure that everyone is asked the same questions. You won't be able to collect the best data for analysis if everyone is asked different questions or only a few people are asked the same questions.

2. You always want to start broad and then narrow down in your data collection. Start by looking at what data can be collected by current databases and then narrow your collection from there. While you are in this process, it is common to think of factors that you didn't think of before. If you want to include something in your data collection, you should always inform other members of your team.

3. Get organized before you start collecting your data. If you plan to place all of your data into files, then get storage for your files. If you are going to set up a computer to keep all your data available on a database, make sure that all the equipment you need is all set up and the database is ready to go.

4. Once you have collected the data, always place it in its correct storage area. This means if you are going to scan the data into the computer, you want to do this right away and save it in the correct location. If you worry about organizing your data after it is all collected, you will quickly find yourself overwhelmed.

Step Four: Analyze the Data

In reality, analysis begins as soon as you start asking questions. You are always thinking about the answer and how to find the answer. Therefore, when you are collecting the data, your mind will continue to piece together the information. By the time you are ready for step four, you will already have an idea of what the data is telling you. This doesn't mean you should skip step four or rush through it. You want to take your time on this step to make sure that you don't miss anything. You want to dig as deep as possible into your data to come up with the best solution.

You want to find the mean, standard, maximum, and minimum deviation of your data. You might find that you have exactly what you need to answer your question as you analyze. However, it is more likely that you will need to go back and reframe your question. It is often common that you need to look for more information. This is why it is important to take your time as you are analyzing the data. The deeper you go, the more you will find what you are missing right away.

There are many analysis tools available to use and it is always a good idea to look into the best tools for your analysis before you start this process. You might find software that helps you digest the numbers, or you might find software to help you with the formulas.

Step Five: Interpret the Results of the Data

Some people feel that step four and step five are the same. Once you use the Lean Six Sigma methods and follow these steps, you will quickly realize that they are different, and it is essential that you separate the steps.

One of the most important parts about interpreting the results is you fail to reject the hypothesis. You don't aim to prove that it is true. You always want to keep in mind that chance can interfere with your results and cause you to reject the hypothesis. It is important to note that you cannot take this personally. Rejecting the hypothesis happens from time to time.

To help you interpret your results, you can ask yourself a series of questions. For example, you can ask yourself, "How does the data answer my original question?" You want to remember your original question and any question you rephrased through the process. Another question to ask is, "Are there any areas that I have not considered?"

4

CASE STUDY: HEALTHCARE

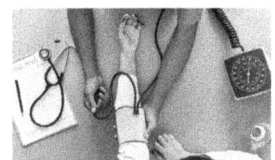

Healthcare is a growing field which is highly interested in the Lean Six Sigma program. Hospitals and clinics from all over the world are working with Lean Six Sigma professionals or paying for their own staff to complete the certification. Many hospitals have a whole team of Six Sigma professionals that consists of hospital stakeholders, leaders, Board of Directors, surgeons, nurses, receptionists, and other staff necessary to run a hospital. Together, they work to ensure that the practices and policies are conducted throughout the hospital. They also come together to focus on ways they can improve the hospital and work closer with other local hospitals and clinics.

While people know that Lean Six Sigma is successful, its practices in many professions is newer. One of the strongest areas that is seeing some of the most benefits is the healthcare field. However, before we go into detail about the case study, it is important that you understand the application is not fully explored.

Hospitals and other healthcare practices are looking into

Lean Six Sigma because it helps organizations achieve goals. Not only can they focus more on helping their patients through Lean Six Sigma professionals, but they also eliminate inconsistent processes that causes them to lose money.

By following the five steps (define, measure, analyze, improve, and control) of Six Sigma, improvements in patient care rise in a number of ways (A Look at Six Sigma's Increasing Role in Improving Healthcare, 2019):

- By improving the turnaround time for labs
- Reducing the number of mistakes made by nurses, technicians, and physicians
- Speeding up the insurance claims process
- Lowering the amount of time people spend in the waiting room
- Improving the outcome for patients

Examples of Successful Healthcare Six Sigma Improvements

The Board of Directors of Rapides Regional Medical Center decided to hire a Lean Six Sigma professional to help them decrease their emergency department's wait times, decrease the hospital's spending, and help their providers see more patients quickly—while not declining the patient's care. After incorporating the Six Sigma leader's methods, the hospital found itself successfully saving over $950,000 every year, increasing the number of patients their doctors see, and declining wait times. The surveys and other methods of communication that the hospital uses to analyze how impressed their patients are with the change continues to help the hospital improve its services (A Look at Six Sigma's Increasing Role in Improving Healthcare, 2019).

Stanford Hospital and its clinics needed to find a way to save millions of dollars as costs were becoming too high. Their goal was to lower costs while continuing to provide their patients with the same or better quality care. The hospital's main focus was its coronary artery bypass graft operations process as this is one of the hospital's largest expenses and services. Through their Six Sigma application, Stanford reached a savings of $15 million dollars annually. On top of this, the hospital's mortality rate declined to about 3% and its other costs decreased by 40% (Aminu, n.d.).

The Women and Infants Hospital of Rhode Island wanted to try to improve its embryo transfer program without increasing costs. They incorporated Six Sigma training and methods into their policies and procedures. The outcome of the process improved the hospital's implementation rate by 35% (A Look at Six Sigma's Increasing Role in Improving Healthcare, 2019).

Application

With all of these successful examples, it is no wonder why so many other medical centers and hospitals are looking into hiring Lean Six Sigma professionals. The University of Toledo Medical Center decided to hire their own Lean Six Sigma professionals to help them with their kidney transplant patients. The goal of their process was to help their patients receive better pre-operational processes. There were five main goals for this study:

1. Improve customer satisfaction
2. Decrease or eliminate errors
3. Improve efficiencies
4. Optimize cycle time to 180 days or less
5. Improve the protocol effectiveness and execution
6. Streamline administrative processes
7. Reduce cost

The University of Toledo Medical Center is known for its successful kidney transplants. Over the last few decades, they have hired some of the best physicians to focus on kidney transplants and close to 2,000 transplants have been performed since 1972 (Franchetti, 2015).

Another part of the application was to give Toledo Medical Center patients the best care possible. While the hospital already strived to do this, they knew that bringing in a Lean Six Sigma professional gave them the best chance to make sure they improved in as many areas as possible, including their patient care. Furthermore, by focusing on patient care, the team would ensure that the changes made did not decrease patient care. If you don't focus on all the factors as a Lean Six Sigma professional, you can find yourself ignoring certain aspects that will cause a drop in patient care or in other areas. When this happens, you won't receive the results you expect or reach your goals.

Measure

When measuring this process, the Lean Six Sigma professionals and team focused on the average total time of the preoperational process. This time is measured in days, with a goal of reaching a cycle time of 180 days or less. After analyzing patient records, the team noticed that the average cycle time for patients was 227 days. The team knew that to make Toledo Medical Center one of the best places in the country to perform kidney transplants, they needed to decrease the cycle time without decreasing patient care or physician performance.

Before patients could qualify as part of this study, they had to become a patient of Toledo Medical Center's kidney transplant unit. This means they had to receive a referral to the transplant center because they are in need of a new kidney, be considered a healthy candidate for a transplant, and complete any tests and labs that are required. Some of these labs and tests include stool samples, blood work, EKG test, colonoscopy, tuberculosis tests, pap smears, diabetes test, chest x-ray, and dental clearance. Once all the tests and labs are complete, professionals will review the file for the patient and decide if they make a good candidate. If professionals determine a patient is a good candidate, the patient is then transferred to the hospital's waiting list. Once they are on the waiting list, it can take up to two years for a patient to receive a kidney. This is on top of the nine months that patients already spent on all the labs and tests required.

The leaders and professionals at Toledo Medical Center knew that decreasing the cycle time for kidney transplants, which includes the time they received the referral and their recovery time, would be difficult. Therefore, they contacted a team of Lean Six Sigma professionals. This team consisted of three transplant coordinators, a financial advisor, two surgeons, a psychologist, and two industrial engineers. The main method that the Lean Six Sigma team decided to use was adapted from a couple of sources (Franchetti, 2015):

- *Partnering with Your Transplant Team: The Patient's Guide to Transplantation*
- *Improving Healthcare Quality and Cost with Six Sigma* by Brett E. Trusko, H. James Harrington, Carolyn Pexton, and Praveen Gupta

To do this, the Lean Six Sigma team decided that their customers were the patients, so everything they did would be centered around their customers. They would work along-side their client, the Toledo Medical Center, to make sure that the policies and procedures aligned with their execution plan.

Take a moment to put yourself into this position. Imagine

that you are on the Lean Six Sigma team and are developing an execution plan. You know the main goal is to decrease the number of cycle days for kidney transplant patients. You want to go from a preoperational process time of 227 days to 180 days or less. One of the biggest factors to consider is that you can never predict the future. You never know when a patient will receive a kidney. There are some preoperational factors that cannot be accounted for until they are about to receive the kidney. The average wait time is two years and you want to lower this to a year, if not a matter of months. On top of this, you cannot allow for any factors to be ignored. You need to ensure that all patients receive the best care from the physicians, nurses, and staff at the Toledo Medical Center. After all, the hospital wants to become the best in their country for kidney transplants and to do this, the focus needs to be placed on the patients.

What type of questions will you ask yourself and the team? What steps will be placed under each of your main goals. For instance, how will the Lean Six Sigma team decrease the number of days in a patient's preoperational cycle? Because you are focusing on the patients, you need to know how they feel. Therefore, you have to get into contact with previous patients who have received kidney transplants and patients who are waiting for a transplant. It is always a good idea to go above and beyond when you are focusing on collecting data, so you also want to talk to patients who are in the middle of the preoperational phase. From there, you believe, the team can analyze the data and decide the best measures to take to decrease the number of days.

For this project, there were several input variables:

1. Conflicts in schedules between patients and employees.
2. The availability of employees.

3. The delays, cancellations, and lack of funding from patients.
4. The lack of real-time data for patients.
5. The various data collection procedures used by employees.
6. The lack of communication among employees.

To give patients the best outcome, these variables needed to be considered into the plan. For the best outcome, the employees needed to work with the Lean Six Sigma team and patients to improve the input variables.

This project had one output variable, and this was time. It focused on:

1. The amount of time between the patient's referral from their primary doctor to meeting with a physician at the Toledo Medical Center.
2. The time it took for employees from the Toledo Medical Center to perform all the necessary tests and labs for the patient to qualify for a kidney transplant.

Analyze

The first step the Lean Six Sigma team did was analyze the previous two years of patient data. This proved to be a daunting task because there was no set system for patient analysis. They found that some patients reports weren't even completed. This meant that the team needed to identify these patients and contact them so they could fill in the blanks.

From there, the Lean Six Sigma team focused on the goal. They knew that the process from transferal to qualification for the waitlist had to be 180 days or less. The reason the team came up with the 180 days was based on several factors:

1. **Patient delays.** Unfortunately, patient delays are something that is out of the Lean Six Sigma professionals' control as this often falls on the incoming patient. It is understandable that some patients wait to contact the Toledo Medical Center because realizing that you might need a kidney transplant is an emotional journey. Therefore, the team focused on the factors they could control. They looked at ways the patient's primary care physician could contact Toledo Medical Center and if the center could contact the patient. They also looked at how the Toledo Medical Center delays their own patients when it comes to communication, testing, and labs.

2. **Minimum time required for the tests.** The Toledo Medical Center had a policy that the patient's tests had to be completed within a certain amount of time. The Lean Six Sigma team looked at this policy and wondered if the hospital was giving themselves too much time, which allowed the employees to feel a bit too relaxed about the situation. While all the employees realized the necessity to get the patients in as soon as possible, the team wondered what the true minimum time requirement was.

3. **The importance of bringing patients into the operating room quickly to save their lives.** It is true with any transplant list that some patients are more critical than others. These patients are often placed near the top of the list while patients who aren't as critical are placed at the bottom. The Lean Six Sigma team wondered if there was anything that could change within this list to make sure all patients receive the adequate amount of care within 180 days or less.

After analyzing the history of the last two years, the Lean Six Sigma team focused on 509 patients. In total, 85% or 433 patients waited longer than 180 days to complete the preoperational process. The team found this number dangerously high when it came to the goal.

The next task was to focus on the data analysis that the Lean Six Sigma team would use to decrease the 85%. The Six Sigma tools the team decided to use to finish and support their analysis include:

- **Root cause analysis.** This type of analysis focused on the last five years of the hospital's history with their kidney transplant patients. It gave the Lean Six Sigma team further support in validating their previous findings of 85% of patients waiting longer than 180 days. They divided this analysis into five different categories: (1) the amount of time between the referral and letter sent to patients, (2) the amount of time between the letter and orientation, (3) time between orientation and patient's evaluation, (4) time between evaluation and the transplant committee's meeting, and (5) time between the committee's meeting and the patient being put on the list.
- **The five why analysis.** This analysis is part of the root cause analysis. It is when the Lean Six Sigma team asks themselves five "whys" that help determine the root cause. For example, they will ask, "Why does it take the hospital 40 days to place a patient in the orientation process from the time they receive the letter?"
- **Process capability analysis.** This analysis focuses on how well a process performs by noting how the process matches with the limits. The limits are the goals or what people want to achieve when they

perform the analysis. If the limits are reached, then the process is good. If the limits are not reached, then people need to look at the process to see how they can reach the limits.

- **Confidence intervals.** This is an important part of the analysis process because it gives the true value to the measurement. The Lean Six Sigma team focused on using the confidence level to measure the process time.
- **Work sampling.** Work sampling analysis is sometimes called work study. This focuses on how much time each employee can take to perform certain tasks. For example, how long does it take to get any letters ready to send to the incoming patients? How long does it take to analyze the patient's results, so the committee knows whether to place them on the waitlist or not?
- **Failure mode effects analysis.** This type of analysis is always important because it looks at the failure rate within the process. For example, how can the process of reaching less than 180 days fail? Lean Six Sigma professionals often use this type of analysis to walk them through the improvement phase.
- **Pareto analysis.** This analysis focuses on the 80/20 rule, which tells people that they can do 20% of the work and receive 80% of the benefits. This analysis is often part of the five whys as it will help people discover where they can take eliminate work and what work is most important when it comes to reaching their goals.

Analysis can take a while when it comes to so many factors. You will want to take the analysis on a step by step basis. For instance, if you are measuring with five types of analyses, you will spend time on each type and them combine them in the next step. It is similar to looking at the pieces of the puzzle before you look at the whole puzzle. You need to have a clear understanding of the whole analysis and its smaller pieces before you can move on to improving.

Improvement

When you are a part of such a large Lean Six Sigma team, you need to hold several meetings when you are discussing improvement. The team will break into smaller teams to break down the results further. For example, in a team of 15 members, you will break down into three smaller groups consisting of five members each. If you are looking at 6 different analyses, each group can look at two types. Once each team has the results, everyone will become a part of the larger group and share the results and how each group came to this conclusion. From there, the larger group will look at each analysis and confirm what the smaller group found.

When it came to look at improvement for the Toledo Medical Center analysis, the Lean Six Sigma team not only created smaller teams but also included patients, stakeholders, and other staff into each team. After several focus sessions were held, the list of official improvements was approved by the team:

1. Look at the logistics of using a video for orientation that the Toledo Medical Center can send to their patients (or have them log-on and view online) instead of needing to meet for the orientation.
2. Once the patient views the video, they will receive the preprocess sheet list. This is usually given at orientation; but, when the patient needs to request this after the video, it is a way to ensure they have watched the orientation video.
3. Develop a patient database to keep all staff informed

of where a patient's progress stands. For example, the database will not know whether the patient has seen the orientation video or if they are on the waitlist.

4. Develop a system for tracking patients and employees to make sure everyone meets the target deadline. This tracking system will include communication between all members involved and testing.

5. Offer more evaluation days throughout the week to speed up the evaluation process.

6. Eliminate the coordinator from every clinical evaluation the patient has throughout the process. Instead, the coordinator is only needed at the end of the process, which is the patient's last meeting before they receive the operation. This will open up time for the coordinators to focus on other tasks that will speed up the preoperational process. Without the coordinator, the patient will meet with other physicians and their case worker.

7. Once the patient has completed the evaluation process, they will receive a process handout that will explain their timeline.

8. Continue to monitor the patient and the database while the team is evaluating the patient for the waitlist.

9. Develop a system to ensure that patients and physicians are on the same page when it comes to scheduling after the evaluation process. This system will also include the potential physicians and nurses the patient will have throughout the procedure.

Control

In this case, control refers to how the process will continue to be monitored even after everything is in place and running smoothly. This is part of the execution plan but is often one of the last parts considered. It is important that people focus on how to control their plan from the beginning. For example, when you are meeting with your team of Lean Six Sigma professionals, you want to discuss control right after you discuss other parts of the execution plan. This helps near the end of the process because you understand the next steps of the plan without having to worry about developing the final stages of the plan.

The Lean Six Sigma team decided that the progress needed continuous monitoring. This means that the databases will continue to be used and staff will collect information from this database about the physicians, other staff, and the patients to evaluate how well the process flows. They will compare the new data to the previous data as this will allow them to look at the bigger picture. They will notice where improvements are met and where there still needs to be improvements.

Employees that are part of the Lean Six Sigma team will hold meetings and report back to the Lean Six Sigma professionals. If there are other improvements that need to be made, the whole team will conduct this business together.

5

CASE STUDY: RECYCLING FACILITIES AND WASTE MANAGEMENT

A nother popular topic around the world is recycling. It seems that small and large communities strive to have their own recycling facility. However, this can be a challenge for many communities because of the cost to provide recycling services. Therefore, many communities have turned to Lean Six Sigma methods to help improve its recycling facilities and decrease costs.

One of the biggest benefits of using Lean Six Sigma to help with waste management and recycling facilities is helping the world. While many people feel that recycling and waste management only help at the local level, it is proven that waste management helps underprivileged children in schools as they receive snacks for free (College Students Use Six Sigma to Solve Food Waste, Recycling Challenges, 2017).

A few students from Rose-Hulman Institute of Technology in Terre Haute, Indiana, focused on using Lean Six Sigma to help their college reduce wasted food. The students did this as a class under the college's mathematics professor Diane Evans. Dr. Evans, who is a Black Belt in Lean Six Sigma understood that many people do not understand the true nature of Lean Six Sigma. Several years ago, as Dr. Evans was teaching her class about the benefits of Lean Six Sigma, she had an idea to give her students hands-on experience. However, she also knew that the process would take longer than one semester to complete. Therefore, Dr. Evans formed a plan where the process of eliminating waste would continue from class to class.

While Dr. Evans knew this work could get dirty, her students were surprised by the real-world statistics and how dirty the work became. The method they used to measure the amount of food wasted was placing this food into plastic bags and then weighing the bags. The results showed that 20 pounds of food is wasted for every 120 students on campus. From there, the students worked together with professors and kitchen staff to come up with a solution that will use smaller utensils when serving food, pre-dishing food in reducing quantities, and limiting the number of glasses and plates that students can take to their table. On top of this, the college decided to get rid of trays, which limited students to only take what they could carry to their table (College Students

Use Six Sigma to Solve Food Waste, Recycling Challenges, 2017).

Once the students reached a conclusion for waste management, they looked into recycling efforts at the college. They decided that another way to reduce waste is to recycle more items than people threw away. They look at factors such as how many recycling bins are available on campus and the students' awareness of recycling.

There were many reasons why Dr. Evans wanted to bring her students into the real world when it came to Lean Six Sigma. For instance, she wanted her students to understand, first hand, what Lean Six Sigma is all about. Through their real world experience, Dr. Evans knew that they would not only read about how Lean Six Sigma can help certain processes and change the world, but they can make themselves a part of this process. Therefore, along with an educational value, the students will also learn real life values.

Dr. Evans and her students at Rose-Hulman Institute of Technology is only one of the many successful Lean Six Sigma recycling and waste management studies. It is something that Dr. Evans continues to manage along with her class as this is one of the most important steps of Lean Six Sigma.

Recycling and waste management go hand in hand. In a way, you cannot have one without the other. No matter how hard people push to start recycling, it seems that most Americans are still set on throwing items and food away because it is easier than focusing on recycling. In fact, recent studies show that about 25% of Americans don't recycle at all (Goodman, 2019). Some people feel it is easier to let the garbage men take everything away and allow the process to go to the local landfill. Unfortunately, this is causing most American landfills to become overfull. Many landfills are unable to contain the items brought in, especially plastic bags. When it comes to these items, they are caught in the wind and start to litter the surrounding area. Livestock and wildlife are often caught in plastic or become sick if they try to eat the plastic.

One of the best ways to get people to see that recycling is the way to go is to understand the basics that Lean Six Sigma professionals will use when focusing on their waste management task—the benefits of recycling. Lean Six Sigma professionals understand that one of the best ways to get people to remain focused on the task is to discuss the benefits.

Recycling Is Beneficial for Everyone

It doesn't matter who you are or where you live, recycling will help you. Waste management keeps food in the stores so many people can purchase the items they need. Recycling also allows people to benefit with different items that are created from recycled material. Another bonus from this is that these items are also cheaper, which helps people who live in poverty. Another reason recycling is good for people and the world is because it conserves energy, decreases pollution, and slows down global warming.

Conserves Natural Resources and Saves Energy

Recycling saves energy and conserves natural resources. When factories make raw materials, it takes more energy. However, when they make recycled materials, it takes less energy. For instance, when paper comes from recycled paper, it uses 40% less energy and recycling a glass bottle can give power to a 100-watt light bulb for up to four hours (9 benefits of recycling | Friends of the Earth, 2018).

Recycling conserves natural resources by saving the forests and trees from being cut down and used as paper. While most people state that we can solve this problem by planting new trees, there is a lot of history when it comes to our natural forests and once these are lost, they are gone for good. Furthermore, wildlife can migrate to other areas or die because they no longer have the natural resources they used for centuries.

Recycling Creates More Jobs

Recycling creates jobs because it is labor intensive. Any factories that handle recycling understand that mixing up the colored paper with the white paper can cause problems with recycled products. Therefore, it is extremely important to make sure that facilities have enough employees to take care of any sorting. On top of this, factories also need help with collecting recycled products. Then, there are other facility operations to take into account, such as logistics, sales, and daily operations. Recycling is like any other business in the world—you have to have enough people to make sure every job is done effectively and efficiently. In 2015, the United States scrap industry created over 320,000 indirect jobs and about 150,000 direct jobs (Leblanc, 2019).

People Begin to Think Beyond Their Walls

Recycling and waste management helps people think beyond their walls. They start focusing on the people around them and other countries. People naturally want to do everything they can to help people live their best life. Waste management is a global problem that causes our water to become unclean, wildlife to become sick and die, and even deformities as there are chemicals in waste management.

When we start to think beyond our walls, we start to see how we are truly treating our earth. We notice that the waste is taking over, and people don't know where to put the excess waste. We pay more attention to the environmental damage waste causes and how people are becoming sick due to pollution and animals are losing their homes because of global warming.

When people start to think beyond their walls, they want to become involved in recycling. They want to learn more about it so they can help people and animals all over the world. Therefore, recycling helps bring people together. It not only unites people in a community, but all over the world.

It Doesn't Cost You Anything but Can Save You Money

Most communities have a recycling center where you can bring your items to them for free. They will have bins set up and all you have to do is put the items into the correct bins. Other communities will give you recycling bins and come to pick up the recycling once a week or so. The most that recycling will cost you is your time to sort the materials into the correct bins. Of course, this all depends on how much you want to recycle. For instance, you can recycle cans, glass, and paper or just choose to recycle paper. But, this means you will need to take time to sort any advertising from newspapers and white and colored paper as well.

Reduces Landfill Waste

While each landfill has a process to eliminate the waste brought in, there is so much they cannot keep up. On top of this, people throw away items, such as batteries, that can damage the natural environment, causing wildlife to become sick, and dangerous for humans. Furthermore, the process that most landfills use to get rid of waste is by burning everything. This contributes to climate change and causes pollution.

Application

In Toledo, Ohio, a recycling facility conducted a study with

Lean Six Sigma principles. The facility, operated by the government, wanted to conduct a study so they could understand how Lean Six Sigma can help improve recycling in the United States. While recycling is a growing industry, it also costs the government a lot of money. This can cause problems, especially later, if they do not gain control over the situation. Furthermore, the government wanted to look at what errors the recycling and waste management industry followed that they didn't realize. In general, the goal was to make the recycling industry more successful in and outside of government controlled operations.

Lean Six Sigma professionals focused mainly on the issues of quality and cost that the recycling facility experienced. The government had to cut costs in the area of recycling because the budget they received was lower than previous years and they understood this budget could get smaller over time. At the same time, the government wanted to look at how they could build on the community's recycling efforts. Was there anything they could do to get more people to recycle?

One of the first tasks the Lean Six Sigma team did was meet with the government officials in charge of the operation and the study. They needed to get a better understanding of the operations history and what the government wanted to achieve. In the end, everyone knew that the recycling facility would need additional equipment to include nearby communities that didn't have any recycling processes. The team also decided that their goal was to become more productive so they could serve the public better and utilize the tax dollars more efficiently. Simply, the team wanted to manage the waste, increase production, and reduce any issues that the recycling facility had at the start of the study.

From there, the team reassembled to make sure everyone was on the same page about the goal, establish a start date, end date, and designate the team and the smaller teams. The project team included everyone working on the project. They broke down into smaller teams as this allowed each team to focus on specific factors of improvement.

Measure

The main type of analysis the team used was Pareto analysis, which is a type of vertical bar graph that prioritizes problems and improvements in a specific order. Other measurement tools the team used were surveys, check lists, time studies, and interviews.

Analyze

Pareto analysis helped the Lean Six Sigma team understand what the top problems were for the recycling company. Through the analysis, they could also find out where the problems originated from. Like most companies, most of the problems came from a few areas of the process. It followed the rule where about 80% of the problems stem from 20% of the processes. Once the team noticed this, they could focus more on the immediate issues and spend less time on other problems. For example, through Pareto analysis, the team learned that the company processed over 11,000 tons of recyclable material every year. Out of this 11,000 tons, the top three categorized materials were:

1. Newspapers
2. Mixed office paper
3. Aluminum cans and corrugated containers

To analyze the time study, the team used a stopwatch to note how long it took an employee to complete their job. From there, the time recorded of the employee and the standard time for the task was calculated. It is important to note that the employee was recorded several times as this could give the team a better average for the employee. The time was processed by looking at the minutes and seconds. For example, over the course of ten sessions, the employee baled mixed office paper an average of 5.3 minutes. However, the standard time is 2.4 minutes. The team then concluded that the steps for bailing or better training could help improve the employees time, so their average is closer to the standard average time. The team conducted time studies for each area in the recycling facility.

After the team conducted all the studies and other data, they begin the analysis phase by using a cause and effect diagram that laid out the main problem and all the causes of this problem. For example, the main problem was the variation in processing the recycled materials. The four main causes were methods, employees, machinery, and material. Under methods, the subcauses were limited quality control and lack of standard operating procedures. Under employees, the subcauses were small staff and poor maintenance. Under machinery, the subcause was the machine often broke down. The subcause for the material was the paper was shredded, items were wet or oily, and the glass was broken.

Another factor the team analyzed was cost. They wanted to gain an idea of the average cost for maintenance and other expenses. The team needed an idea of where most of the money was going so they could focus on creating a more standard budget that would fit everyone's needs. This budget included all labor expenses and any type of maintenance on the machines and facility. With over a million dollars placed into the facility every year for salaries, operating expenses, and other means, the Lean Six Sigma team had a baseline for their improvements.

The Lean Six Sigma team analyzed five sources of waste throughout this phase:

1. **Transportation.** While the recycling facility had strong processes, they could use improvements. The Lean Six Sigma team noticed the amount of time one product, such as mixed office paper, was moved from one area to the next. This is seen as a waste of time, which causes the facility to lose money. It also makes some employees waste time as they are focusing on sorting or transporting material from one area to the next. The goal then became to bring the materials to its rightful zone when they are brought into the facility.

2. **Defects.** The team clumped the defective materials and the machines into one category. Any type of defect takes away time when it comes to processing. For example, if the machine breaks down and needs a new belt, the employees need to wait for the machine to get fixed. The Lean Six Sigma team also noted that many employees fixed the machine themselves so they can continue working. Unfortunately, they didn't have the right specifications to repair the machine correctly. While this saved money as the company didn't need to call in an engineer, it cost them money in the long run as it took time away from the employee's job and caused machines to breakdown regularly. In the time study, the Lean Six Sigma team noted close to 30% of time was wasted due to defects (Franchetti, 2015).

3. **Motion.** Motion included the time employees took to transport, stack, and handle materials. The Lean Six Sigma team realized that most employees took too much time with the motion and they considered it a waste of time for processing.

4. **People.** Throughout their analysis, the Lean Six Sigma team noted that many employees had to stop their job in order to clean up or take care of a task

that was not included in their list of duties. For example, a trained machine operator needed to stop their job to sweep and mop the floors or transport materials. Because these responsibilities were not part of the employee's job description and took time away from their main duties, it was considered a waste of time. Instead, cleaning and moving materials needs to be designated to another employee where this is their set job.

5. **Waiting.** Another factor that the Lean Six Sigma professionals noted was the wait time. They saw many employees waiting for material delivery, instructions, or for equipment to get set up, cleaned, or repaired. In fact, some cycles showed close to 50% wait time, which included added breaks, bathroom breaks, and machines breaking down.

Improvements

Once the Lean Six Sigma team finished the analysis phase, they moved on to ways they could improve the recycling facility. One of the first points the team noticed was the amount of money the facility lost when it came to their recycled products because of various defects, such as broken bottles. When it came to glass, the team realized the facility lost nearly $800 every year. In total, the facility lost almost $1,000 and over 30 tons from defective materials (Franchetti, 2015). When the team decided to further analyze these findings, they noted that most of the cost came from loose shredded paper, time wasted in transporting materials to the facility, and broken glass.

As a way to try to improve their results, the Lean Six Sigma team looked at eliminating the waste, including time. The three factors they looked at when considering improvements were:

1. **Better instructions and information for the employees.** One of the biggest things the team noticed was the employees didn't always understand waste management, how much their job meant beyond the walls of the facility, and didn't receive the best training. Some areas lacked consistency and flow, which caused the processes to slow down. When one area within the facility slowed down, the rest of the areas followed. To add more value, the Lean Six Sigma team looked at better training procedures that included an educational component, so employees understood their job and its importance. The employees also learned what materials were recyclable and which ones were not. This allowed the processes and the machines to run smoothly. Other instruction included telling employees what they could do while they were waiting. The Lean Six Sigma professionals met with other team members and asked them to brainstorm other tasks for these employees. They asked themselves if employees could perform a different task while they were waiting or help a neighbor with their task, so everything is completed efficiently.

2. **Materials that are free from defects.** Not only did employees start to notice when the materials were defective and set them aside instead of in the recycling machines, but processes were placed to keep employees from getting damaged materials. The process stated that all materials sent to the employees and machines had to be clear from any defects.

3. **The capacity of machines.** The machines used needed upgrades and maintenance so they could handle the amount of processed materials without breaking down often. Some machines received continuous repairs of the same part, such as a belt, so

they could perform the work. The Lean Six Sigma team realized that much of the repairs were temporary as no one looked at the base of the problem. To add value to the machines, the new policies and procedures stated that all machines had to be kept up-to-date on repairs and when one broke down, the employees needed to find the root of the problem and notify the appropriate party for its repair. A lot of money is wasted with temporary fixes, so one small goal was to eliminate temporary repairs.

Control

Once the team finished the analysis and improvement phase, they focused on their control plan. They needed a plan in place that allowed the facility to continue monitoring the improvements. To handle this phase, the team decided to use documentation, control charts, and make sure that the employees continued to communicate with each other, and everyone understood their processes.

Another part of the control phase is to analyze the improvements and take another step into verifying that the improvements are correct. Verification is important in statistical measurements and analysis because it gives teams the added benefit of knowing everything is truly running smoothly and there was no mix up when it came to the first measurement.

Another part of the control focuses on employee training and continuous training and education about recycling and waste management. It is easy for employees to get into a system where they don't focus as much as they did during training. This happens naturally as people become more comfortable and used to their tasks. When people perform the same tasks several days out of the week, or several times throughout the day, they can become mindless and make more mistakes as they are not fully aware of their actions. While mistakes can happen, they are also a waste of time. Therefore, employees are more aware of their actions and the tasks they are performing as an effort to help increase productivity.

6

CASE STUDY: EDUCATION

The importance of education cannot be stressed enough. Unfortunately, most students complain that they don't have the best experience in school. It doesn't matter if they are in elementary, jr. high, high school, or college. Their experience depends on what the school system, teachers, staff, and students are willing to put forth into the experience. However, when students feel that the leaders of the school don't care to put in enough effort to make the school a safe and comfortable environment for them to learn, why should they? Students, especially younger students, learn from their teachers and school staff. Therefore, it is always important for schools to focus on what they are teaching their students and how their students are learning.

There have been many cases of children dropping out of high school because they feel they are too far behind and the teachers don't have the resources to help the student succeed. Children start to fall behind in their classes in elementary school because the classes are too big for the teachers to handle. The schools don't receive enough funding to hire assistants for the teachers to help manage the classrooms. This also creates problems as many students will put more effort into causing trouble in the classroom than learning. Universities are also starting to suffer in this way because they don't have the funding to hire more professors to teach the classes students need for their majors. This causes students to enroll in another year in college, if they can afford it. Even if they can't, more students are finding themselves in heavy debt because of their student loans.

One area in the United States that seems to be on the decline is the educational system, especially in public schools. Over the last few decades, the education system has lost thousands of dollars in funding. Today, teachers all over America are dipping into their pockets to provide their students with enough supplies. The hot lunch program is often in the news with schools trying to find ways to help children receive a hot lunch but also receive the money to pay for lunch. With the declining state of the public school system, many schools are turning to Lean Six Sigma methodologies to help improve the state of their school and services to its students. Directors, superintendents, principals, teachers, and other staff members want all their students to have the best experience. This means they have to work together when it comes to incorporating Six Sigma practices.

For years, many people did not believe that Lean Six Sigma practices could be used in education. However, people are now aware of the importance of these practices and how they can help improve education in several ways.

- Lean Six Sigma and education go together because they are both focused on the customer, which is the students in education. When schools focus more on their students, the policies and procedures become more compatible and students enjoy going to school.

Knowing what the students need, the schools are focused on helping their students grow and aim to bring better services to their students through governmental and state funding and grants. Better services mean students will pay more attention to their studies and strive to get good grades, which aids schools in getting adequate funding.

- As the school receives better funding and services, students are expected to do more when it comes to their classes. This means students are studying harder and are more prepared for college.
- A better school system increases competition with other schools on a national level. When students are competing with other top schools, they put in more effort because they want to remain on top. This helps them excel academically.

It is important to note that education has different areas. Each area requires its own Six Sigma practices. This can bring in an extra challenge for the Lean Six Sigma professionals as they have more than one area to focus on. Because of this, it is a good idea to have a strong team of Lean Six Sigma professionals. Some teams will have a Six Sigma leader per area while other teams will have more than one professional focusing on the system as a whole. The three main areas in education that Lean Six Sigma practices focus on include:

1. **Enrollment.** The only way education can take place in a school is if students are registered and go to their classes. One of the struggles that many schools face, especially high schools, is the enrollment process. While the students sign up for their classes, the administration takes care of the rest. Students also need to have a number of credits and a certain grade

point average (GPA) to graduate and go off to college. This means that the administration needs to make sure every student can take the classes that they need in order to graduate.

2. **Administration.** The people behind the scenes that make sure everything runs as smoothly as possible for the students is the administration. The members of the administration, from the superintendent to the janitor, need to make sure that everything is ready for the students when they come to their classroom. The teacher needs to prepare the lesson plans and ensure each student understands what they are teaching. The administration team needs to ensure that the policies and procedures are in place and everyone is doing what they need to do. They also need to make sure they catch any breakdown in communication as this is an essential part when it comes to the administration team.

3. **Academics.** Academics goes beyond the material students need to learn in school. It also focuses on the teachers and that they are following state regulations. Lean Six Sigma professionals will help teachers learn the best procedures to make sure everything is up-to-date. The professionals will also help the school by hiring the right teachers as students are more likely to learn if they are comfortable with their teacher. The quality of learning within the classroom will strengthen when it comes to Lean Six Sigma practices.

Currently, Lean Six Sigma and education is quickly growing. In fact, there are many schools that are starting to not only incorporate Six Sigma practices into their schools, but also include training for their juniors and seniors as they believe this will help them advance in the path towards college.

There are also several colleges that are including Lean Six Sigma into their course schedule. These classes are taught by Lean Six Sigma professionals.

Below are a few examples of schools that have incorporated Lean Six Sigma policies into their procedures (Districts Turn to Process Improvement for Better Student Outcomes, 2018):

- The state of New York has established a five year plan for student achievement. They want to raise their high school graduation rates and help more students become prepared for college. The goals not only focus on student achievements, but also focus on the continuous process to continue to record the achievements students reach after the five year period.
- The Des Moines School District in Iowa incorporated Lean Six Sigma practices into their school when they established the Department of Continuous Improvement. The goal of this department focused on saving money on textbooks while making sure all the textbooks are up-to-date. They also focus on recycling and improving student experiences, so they are well prepared to enter college.
- The state of Tennessee also incorporated Lean Six Sigma practices to improve their schools across the state. They created a School Improvement Support Network that focuses on each district. Within each district, the group will go to the lower performing schools and help them establish plans to bring their student performance up to standards. Members of the support network group are trained to go into the school and train the teachers and staff in the areas

where they struggle the most. They also help the schools find the root cause of their problems and develop a plan to overcome this problem. On top of this, each school devises their own continuous collection data method to ensure they are on the right path each school year and continue to improve.

- New Mexico started focusing on tracking their student progress more at an individual level. For example, the database focuses on which students are behind on their credits, evaluates the student's grades, and looks at how many students are transferring schools and why. Through this data, the school can find the root cause of students transferring, why students are not earning enough credits to graduate on time, and why students do not receive the high grades they are capable of.

A midwestern university in the United States decided to use Lean Six Sigma practices as a way to help their students enjoy their college experience better. Many students complained that the advising and services they received were poor. Alumni of the university stated that they weren't fully prepared for their career because they didn't receive the education they paid for. Over time, the college's enrollment started to decline. This caused a snowball effect within the budget as most of the university's funding focused on student tuition. As a way to change their methods, the provost of the university contacted student services and asked them to look into Lean Six Sigma methods to improve the situation. From there, student services contacted faculty in the business department. Together, they looked into hiring Lean Six Sigma professionals to help improve student satisfaction and increase student enrollment.

Once the Lean Six Sigma professionals came on board, they noted that the first factor the university needed to realize was that the students were their customers. The university went years without asking students what they thought of the services they had to offer, helping students when help was requested, and didn't have adequate services for their students. Some of the students didn't use the services provided by the college because of the conditions surrounding the service. For instance, students struggled to find the right books for their reports and papers in the library. Therefore, they often went to the public library or used the Interlibrary Loan service.

On top of being the customer, the students were also the product. This gave Lean Six Sigma professionals an extra challenge as they needed to distinguish where the students were the customers and where they were the product. After thorough observation and analysis, the professionals noted that students were the customers of the student services center and the product when it came to the educational activities.

Application

Once the Lean Six Sigma team knew where to place the students, they looked to create the best team for the project. Included in the Lean Six Sigma team were employees from the college administration, student services groups, and students from various focus groups. Over the course of several meetings, the team learned why enrollment was decreasing. Not only did the students have to deal with poor classroom structure, but they also dealt with rude comments that were sometimes unprofessional from student services and long wait times in student services. Just like any business, students talk about their experience in college. If incoming students learn a university has poor student services, they will look elsewhere. Students are paying heavily for their education and they want to ensure that they are given the best education and college experience.

The Lean Six Sigma team decided that their goal was to improve student enrollment by 5% within the first year, increase student improvement ratings by 15%, and decrease the amount of time students waited for services, such as advising, by 25% (Franchetti, 2015).

Measure

The first step to measure student satisfaction in real time was to give the students a survey. Student services conducted this survey through email and allowed students to complete the survey online for easy access. Knowing that students sometimes need a push to complete a survey, student services announced a drawing that would be held for everyone who completed the survey. Three students would be awarded one of the computer tables as a prize.

The survey asked the students 15 questions and asked them to rate their answers on a scale of 1 to 5 with 1 being "very dissatisfied," 2 being "somewhat dissatisfied," 3 being "neutral," 4 being "somewhat satisfied," and 5 being "very satisfied" (Franchetti, 2015).

The questions the students had to rate, and their average ratings are as follows (Franchetti, 2015):

. . .

Question explanation

Average rating

How satisfied they are with their overall college experience at their university.

4.2

If they feel like they belong at the university.

4.8

The quality of their general course experience on campus.

3.1

The availability of the courses they need for their major.

4.5

Their satisfaction with the registration process.

2.4

The quality they receive from academic advising.

2.0

If they have an organized experience outside of the classroom.

5.1

Their informal experiences outside of the classroom.

4.9

Their satisfaction for online advising services and its availability.

1.6

How safe and secure they felt at the university.

4.7

Their satisfaction with the quality of their major courses.

3.9

Their satisfaction with their advisor's availability.

1.4

Their satisfaction for the referrals and information they receive from their academic advisor.

1.7

Their satisfaction with the orientation program.

2.8

They had to answer "yes" or "no" to if they would return to that university if they started college over again.

Yes 58%

Analyze

Once the results from the survey came in, the Lean Six Sigma team tallied all the results and then rated the questions. They focused on the ratings of questions that were lower, such as online advising and accessibility of advisors, before focusing on higher rated questions, such as organization outside of class and safety on campus. With this method, the top three concerns that the team needed to focus on immediately became:

1. Availability of online resources and advising
2. Availability of advisors
3. Information and referrals received from advisors

The Lean Six Sigma team combined the survey results with the time study data that showed students waited an average of 7 minutes to meet with their advisors. Furthermore, the time they spoke with their advisor averaged around 22 minutes. This meant that the Lean Six Sigma team wanted to focus on lowering the wait time. They also told the student focus team members to look into finding ways the advisors can help their students more quickly. For example, decreasing the 22 minutes to 15 minutes. The Lean Six Sigma team wanted to know if the advisors are wasting time when they are talking with their students or if they are organized before the students walk in and the advisors try to help them as quickly as possible.

Other analysis that the Lean Six Sigma team looked at were the results the student focus group noted from the survey. The student focus group stated that students wanted advisors that were specific to their major. For instance, a history major didn't want an academic advisor from the business department. Another way that students believed would improve their academic advising experience was accommodating the advisor's hours to the student's schedules. Advising hours were only from 9:00 A.M. to 3:00 P.M. Students felt since most classes went until 5:00 P.M. that advisors should be available for after hours, such as after 5:00 P.M.

Another concern was the communication between the students and their professors and advisors. Students complained that they regularly sent an email to their professor or advisor with a question or concern and they wouldn't always receive a response.

Improvement

Working closely with the student focus groups and student services, the Lean Six Sigma team decided that the university could improve the students' experience in many ways. These improvements would make sure that more students wanted to come to the university, helping the college receive their increase in enrollment. Furthermore, the overall experience of the students will improve, making more students say they would come to this college if they were starting over. The Six Sigma team also believed these improvements would decrease the transfer rate from the university, creating a higher graduation success rate. This is important to many incoming students as they regularly look at the university's graduation success rate when choosing a college to attend.

One of the improvements to the student service processes became adding an online chat and text opinions into the advising procedures. This allowed students to contact their advisors between their classes for any questions or concerns they had. They could also set up a meeting with their advisor so the students wouldn't have to deal with the wait times and the advisor can prepare for the student's visit.

Another improvement became extending advising hours to 7:30 P.M. from Monday to Thursday. This allowed students to meet with their advisors face to face after their classes. On top of this, the advisor's students received were part of their program. The university hired another person to help with academic advising for each major they offered. This person also helped with any technical and job prospect questions students had. Advisors didn't always have to be at their desk for meetings. They could meet with the students online or through text message.

Developing an online database that recorded student's records was another improvement. This database allowed advisors to look at student records electronically from any computer through a log-in system. This meant that advisors could access student records in and out of their office, depending on where they were holding advising hours.

Control

One of the biggest methods of control the Lean Six Sigma team decided to implement was student surveys. Twice a semester, student services emailed students a survey that asked them to rate their experience at the university. On top of this, tracking tools were used to analyze the continuous improvement of the university. Every advisor and professor are given the results of the study and informed at how the university is improving.

7

APPLYING SIX SIGMA TO YOUR START UP BUSINESS

E very day, more and more people start their own business. With the Internet, people often turn to an online business that gives them the benefit of working from home. There is a lot of effort that goes into establishing your business and ensuring its success. One of the strongest services people are using for their business is Six Sigma methods. While some business owners take the steps to receive Green Belt training, other people contact a Six Sigma professional to help them improve their business.

It's not easy to start a business. However, when you have the drive and will-power to succeed, nothing can stop you. There are a lot of steps that go into establishing a successful business—and keep it going. For instance, you need to start by creating a business plan. This plan lays out your goals, mission, and process you will use to start your business. It will discuss the type of services you need, your target customer, roles for your employees, and how to make sure everyone is on the same page and pleased with the business.

Usually, one of the last steps people are thinking about when they are starting up a business is Lean Six Sigma policies.

They feel that this training is for companies that are already successful. However, one of the best ways to start your business off right is to incorporate Lean Six Sigma policies immediately.

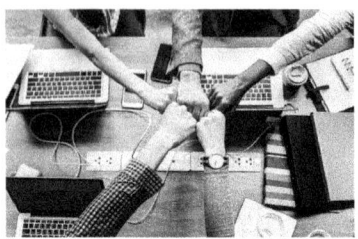

Benefits of Incorporating Lean Six Sigma into Your Business

Lean Six Sigma improves all areas of a business as long as the practices are understood and used every day. Some of the strongest benefits of bringing Lean Six Sigma into your business include:

- **Improve your bottom line.** No matter how or where you advertise, the best references will come from customers who are satisfied with your business. They will tell their friends, family, co-workers, and even talk about your business on social media. Happy customers always return to your business and they will usually bring a friend. This will improve your revenue, which helps make sure your doors stay open and allows for your company to grow.
- **Establish your target customer.** Every business, no matter how large or small, needs to focus on their target customer. A target customer is necessary because people are individuals. If you try to focus on every customer that walks through your door, you will find yourself overwhelmed by your customers'

needs and how to improve your products. But, when you focus on your target customer, you are grouping your main customer qualities into one and allowing yourself to improve your business and keep your customers satisfied. Think of the target customer as your most loyal customer who is willing to go above and beyond to help your business succeed.

- **Ensure your employees are satisfied.** It is easy for businesses to put so much emphasis on making their customers happy that they forget what their employees need. While many business owners find the customer and employee balance a bit tough to find, once you reach a strong environment that ensures your employees and customers are satisfied, you are guaranteed to have a successful business.

- **Keep everyone on the same page.** Communication is key when it comes to a business. Whether you are communicating with your customers or employees, it is essential that everyone stays on the same page. It is common to find a breakdown in communication through the process. There are times where you won't realize the breakdown in communication and other times when it becomes noticeable. Lean Six Sigma can help keep everyone on the same page because you become more mindful of your practices and follow procedures to improve communication between the company and with the customers.

- **Start the monitoring process from the beginning.** As you are developing your business and its practices, it is hard to know what monitoring process you will need. Usually you will learn what areas you need to monitor and what procedures need improvement as time goes on. Unfortunately, this can lead to problems. Lean Six Sigma training can help you establish a monitoring process from the start, so you always have a clear idea on what direction your company is growing.

Grow Your Startup Business the Lean Six Sigma Way

If you are starting your own business, there are a few factors you can incorporate without the need to reach out to a Lean Six Sigma professional. These are also factors that every Lean Six Sigma professional should know when they are getting ready to help a startup business.

1. **Don't waste your time.** Making the most out of your time is a difficult step because people don't always realize how they are wasting time. They think that as long as they are sitting at their desk or spending time working on their task, they are spending their time wisely. In reality, you can be wasting time while you are working at your desk. Think of all the moments you stare into space, look at your social media accounts, or look at a certain part of a project instead of moving on to focus on something else. There are many ways that you can be wasting time and you don't even realize it. Lean Six Sigma professionals are trained to notice moments when you are wasting time, even while you are working. The key is to work efficiently, which is something that most people don't think about while they are working. For instance, do you have any distractions in your area? Are you sitting by a window? Do you have music playing or is the television on for background noise? While you are not giving the background noise your attention, it can slow down your process as you are focusing on your work and the noise.

2. **Leaders should receive Six Sigma training.** If you are interested in Six Sigma training, it is time to take

the step into getting the certification. This will help your company as you will always have the training necessary to make sure your company continues to grow with Six Sigma practices. Six Sigma is a special type of training that goes beyond the research you will find online. Six Sigma uses special tools and techniques that are specific to them. The training gives you a different mindset that allows you to focus on the roles, organization, and management of Six Sigma. If you don't want to receive the belt certification, you can always become a Lean Six Sigma Champion. This is a two-day training session that gives you all the basics and helps you understand what questions to ask.

3. **Evenly spread your resources.** You will have a lot of ideas and projects that you are working on at once. It is important to make sure you have all the people and resources you need to get the projects completed on time. When you establish a timeline and create a plan, you want to do everything you can to make sure you finish your projects in the allotted time frame. If you are starting out and you notice that your expectations of finishing a project were not realistic, then bring yourself back to the drawing board and refocus your plan and timeline. You want to ensure that everything is set so you are following all the steps in the established time frame. This will also help you focus and keep you from wasting time. If you come to a project and you only need three people for it, don't ask four people to work on it thinking the project will get done faster. Instead, you want to take that fourth person and have them work on another project.

4. **Always use a mentoring process.** Even Lean Six Sigma professionals have mentors that help them

when the going gets tough. Whether you are a Black Belt, Green Belt, or a Six Sigma Champion, you want other Six Sigma professionals on your side that can help you learn and thrive in the profession. You might find these mentors while you are in your training or as you are looking for other Six Sigma professionals in your area. Lean Six Sigma professionals are here to help each other, especially as they become more experienced. Even though you might be just starting out with your Six Sigma career now, you will quickly find yourself mentoring other beginners who are entering the field.

5. **Invest in software.** Some people are more comfortable with software than other people. There are a lot of software available for the Lean Six Sigma profession and program that allows you to develop and focus on your project and continuous improvement. This is software that many people consider expensive, but it will most definitely help you. While you can use a datasheet from Excel, there are a lot of software programs that include special tools to help you achieve your continuous improvement goals. These tools are specific to the software, so it is important that you perform extensive research, so you know what software is the best one for you.

6. **Analyze and trust your system.** Your system is the reason you want to make sure you do everything right the first time. If you find yourself in the middle of a project and you need to go back to the drawing board because you didn't think of several factors, you will waste a lot of time. Once everything is started, you need to put all your faith into the plan and system your team developed. If you find yourself questioning the steps, you are wasting time with

your questions and you don't trust the system. To help you make sure the system is right for the company from the beginning, you want to analyze the whole system before you decide to go forth with it.

Application

There is a difference when you are using Lean Six Sigma practices in a small company versus a large company or corporation. When you are using Lean Six Sigma for your startup company, you will focus on different aspects than you would in a larger company, such as your service area, your products, and your target audience. In ways, a smaller company is easier to measure because there aren't as many factors to focus on. At the same time, it can be more difficult because there are less people to help you manage your business and ensure the Six Sigma practices are put to use correctly.

One of the biggest setbacks for many startup companies is they don't have the funding to contact Lean Six Sigma professionals to help them make their business grow quickly and efficiently. Therefore, employees will either look into the Green Belt Six Sigma certification as a start or do their own research and follow the processes they find online. While there is nothing wrong with this, many Lean Six Sigma professionals are starting to take note of this as an opportunity. For instance, if you are starting your own Lean Six Sigma practice, you can give startup companies a discount for your service. This is always a good idea for a Lean Six Sigma professional who is starting out as it gives you experience. However, you always want to ensure that you have a team of Lean Six Sigma experts available to help you handle the difficult situations, walk you through processes, or help you with your questions and concerns.

A startup company contacted a local Lean Six Sigma professional to help them establish goals. The company is a restaurant in a small town and is owned by a couple. They have always dreamed of running their own restaurant and recently got enough money to purchase the building and get their dream going. However, they are starting to find themselves stuck in the process. Afraid that they will need to let their dream go, they decided to look into Lean Six Sigma. While one of the owners thought about getting their own certification, they quickly realized they didn't have time to do this. Therefore, they contacted a Lean Six Sigma professional who offered to help them at a lower cost.

Measure

Before the Lean Six Sigma professional looked at any measurements, they talked to both of the owners, separately and together, about their business plan and process. The professional asked the couple about their target customer and what the goals of the business were. They asked where the couple saw themselves with the restaurant in five years.

Another measurement the Lean Six Sigma professional focused on was looking at the prices on the menus and the food offered. The professional compared the restaurant's menu to local menus.

A third measurement the Lean Six Sigma professional used was talking to community residents about a new restaurant. The professional asked the residents what they thought about the new restaurant and if they could see themselves visiting it.

Finally, the Lean Six Sigma professional looked at how well other restaurants are performing in the area. This allowed the professional to get a smaller picture by focusing on the new restaurant and a larger picture by looking at other restaurant statistics in the area.

Analyze

Through all the research, the Lean Six Sigma professional noted that people weren't generally excited about a new restaurant opening up. Many stated that they saw a lot of restaurants open and go out of business in a short period of time. For a small area, there were a lot of restaurants for residents to choose from and most people didn't think of looking into the newspaper advertisements, which is where the restaurant is being advertised.

One of the biggest downfalls the Lean Six Sigma professional noted was the couple lacked communication when it came to their business. They were not on the same page in real time or in five years. One partner had come up with the business plan but didn't take time to explain it well to the other partner. Neither partner had a thought about the mission for their restaurant—they simply knew that it was their dream and they wanted to run a successful restaurant.

When it came to the menu, the Lean Six Sigma professional noted that the business had unique food compared to other restaurants, but their prices were much higher. Therefore, the professional looked at an average dish that a customer would receive and compared this to the amount of food people received at other restaurants. The professional noted that the food portions were about the same. They knew this would reflect on the restaurant as people would feel they deserve a larger portion of food for the higher price.

When it came to other restaurants in the area, the turnover rate was higher than average. About 30% of restaurants that opened over the last five years closed within its first two years. This meant that it became more important for this restaurant to advertise outside of the local newspaper and show customers that they have something other restaurants don't. The restaurant needed to stand out to give itself a higher chance of surviving the first two years.

Improvement

The first method of improvement the Lean Six Sigma professional focused on was getting the owners on the same page. While they were a married couple, they felt that one could take some of the tasks and the other partner would pick up the rest of the tasks. They didn't communicate about work after hours because they wanted to keep their personal and professional lives separate. The first step the Lean Six Sigma professional brought to them for improvement was to discuss their business regularly. They would set up meetings, whether every day or a couple of times a week and talk about where the business is heading and how well their tasks are performing.

The Lean Six Sigma professional also helped them look into developing a database that would keep anyone they hired on task. This database included job descriptions, when orders need to be placed for the food truck, all of the restaurant's policies, its mission, the goals of the restaurant, and all of the training procedures. The owners also included all of their procedures, such as customer service and register procedures.

For another improvement, the owners developed an advertisement plan that would reach beyond the readers of the newspaper. The restaurant became a part of social media and they came up with keywords and phrases to use that would grab someone's attention and make the restaurant stand out.

In an effort to ensure their prices were fair, the owners did further research and collected data on other restaurants that were similar to their restaurant. They looked at the amount of food the restaurants gave and the prices. Of course, they also looked at the location of the restaurant as small town prices are naturally cheaper than prices in larger cities. The owners decided to offer daily and weekly specials and include larger portions of food instead of lowering prices as this seemed like the most feasible option with the price to purchase food from the truck, other supplies, general operations, and paying their staff.

Control

To control their results, the Lean Six Sigma professional advised the owners to develop a database that would allow them to record their daily sales. They also kept track of their daily customer count and gave customers a quick survey to fill as they paid for their meal. With each survey, the customer would receive a free drink or dessert on their next visit.

8

CONTINUING TO MANAGE SIX SIGMA

One of the biggest mistakes Lean Six Sigma professionals make, especially when they are new to the field, is not thinking about how they will continue to manage the improvements within an organization. Many people feel that once the project comes to an end, this is it and they can move on to their next job. While you can typically move on, you want to make sure that you are continuing to monitor your previous task, so the company doesn't fall back into old habits.

Continuing to manage your project is often easier than the steps you took during your project. Everything is set up; the key is to keep the employees of the company motivated to continue the improvements for the best possible outcome. Fortunately, by the time the project ends, people notice the benefits and they are excited to continue the improvements.

Another reason to continue focusing on how the improvements are progressing is because it will allow you to notice any flaws in your plan or if there are other areas that need improvement. You want to keep the improvements as perfect as possible and this means you will sometimes readjust the processes as you or someone else on the team notices that one step will work better if it's completed a different way. Just because you or someone else notices flaws doesn't mean that the plan was bad. It is normal to notice a few flaws and do what you can to provide a better outcome. It is all part of the process, which makes monitoring your outcome more important.

Benefits of Continuous Improvement

Businesses are quickly picking up on the Lean Six Sigma method because of its focus on continuous improvement. While you might leave the company, it will never stop using the methods you brought into the company. One point of the Six Sigma method is that businesses continue to improve, even when they reach their goals. Instead of letting the company float once the goal is reached, people strive to establish a new goal and work toward creating steps and reaching this goal.

Many people feel that continuous improvement is the "Lean" way. While this requires a lot of work, there are a lot of benefits that come from continuous improvement.

Lowers Turnover of Employees

There is a saying that if a company's turnover is high, it is a bad sign. It is often a sign of poor management, a tough work environment, and unhappy employees. While this isn't always the case, companies that have high turnover rates are usually rated lower. Therefore, companies strive to have low turnover. Not only does this help the company by keeping their employees on board, but it makes the business look better to other people. Furthermore, turnover is expensive. Companies spend a lot of money when they need to recruit and hire. It usually costs money for businesses to train someone new because they need to hire more employees for a period of time to cover extra shifts or there are fees associated with any paperwork or testing.

Training is another reason companies want low turnover rates. In many businesses, it can take anywhere from months to years to cover everything an employee needs to know. This makes training more expensive as there are conferences, recordings, testing, and other online resources the employee needs to complete. If the employee isn't at their location for the job because they are training, this means another employee needs to cover the trainee's shift.

There are many other factors that make businesses strive for low turnover rates. Because of this, they are always looking for ways to lower turnover rates. Fortunately, Lean Six Sigma can help businesses in this category because the methods give the company and its employees a sense of accomplishment. Employees are proud to work at the company and will work hard to make sure the company succeeds.

Employees are More Engaged

Every business owner understands that the happier their employees are, the more engaged they will become. When employees feel that their place of business cares about them, they care about the company. They want to nurture the company and help it blossom.

Increases Customer Service

Customer service can make or break a business. Any business knows that they have to make their customers number one if they want to stay in business. It doesn't matter if they are selling products or run a local restaurant. When employees feel that they are taken care of by the business, they will continue to support that company. But, when employees have a negative experience with the business, they will quickly turn away and will refuse to lend any type of help.

The need to make customer service strong for a business has increased over the last decade because of social media. People are quick to post their negative customer service experiences on Facebook, Twitter, or any other social media platform. Many people will publicly try to shame the company by posting on their Facebook wall or tagging them in their personal post.

Lean Six Sigma helps with customer service because it gives employees an outlet for understanding their customers. When employees feel that they are making the customers happy by performing their work, they will follow through with the procedures and policies. Lean Six Sigma also helps identify customer's values and allows the business to establish goals surrounding these values; creating a better delivery process.

Over time, the continuous efforts from the business will pull more customers in, which gives the company a higher profit. From there, the entity can provide better services and products. It can continue to grow, which will allow the company's customer base to grow as well.

Continues to Help Your Company Grow

Another benefit of continuous improvement is your company will continue to grow. Without growth, your company will come to a standstill and you will end up closing your doors within the next few years. For example, a clothing store that has been open for over 100 years but refuses to set up online shopping will close their doors over a clothing store that sets up online shopping. Today, many people would rather shop online than head to the store.

By using continuous improvement, you will find areas where

your store is struggling. This allows you to catch the problem early and find ways to change it so you can continue to build your company. You are less likely to fall back into your old ways when you focus on constantly changing and improving yourself. You will continue to reduce mistakes and improve the quality of your business internally and externally.

Success Factors

By following the steps of Lean Six Sigma practice, you will help businesses become most successful. However, it is up to them to continue to implement their improvements. One of the most successful factors of the Lean Six Sigma journey will depend on the business' implementation effort.

However, there are several success factors to consider throughout the Lean Six Sigma journey.

Refuse to Accept Defects

The main reason why businesses struggle is because they have internal defects. These defects can be within their product, communication efforts, or their customer service skills. As a Lean Six Sigma professional, you do not want to accommodate or accept any defects. It is up to you to make sure that employees of the company recognize the defects and will use the improvement methods continuously so they can overcome these defects. For example, you are helping a company improve their communication. The lack of communication between the executive team and other employees of the company was the first factor you noticed.

While the team is analyzing data, you realize that most of the employees feel that they are not informed about changes going on with the company because the executive team doesn't tell them what they need to know. There are many times where employees are asked questions about their weekly specials or events happening and they can't give the customer all of the information because they haven't been told the information themselves. This creates a problem with the customers as it doesn't give them the best customer service. Furthermore, customers notice the lack in communication and don't feel the company is organized. Employees are afraid that the company will start to lose their customers because of the lack of communication.

As you bring up the analysis to the executive team during the team meeting, one of the members states that they will start posting information for the employees to read. Immediately, you start to notice defects within this plan. While it is a good start, you want the team to establish a plan that won't have any type of defects. You need to follow the zero tolerance policy for defects and ensure that the executive team develops a plan that will make everything head in the right direction immediately. You continue to follow this method until the plan is in place.

Always Find the Root Cause of the Problem

You will find there are many reasons for one problem. However, it is more important that you find the root cause of the problems as this is the first cause that needs a successful resolution. You will find that fixing most root causes will fix other causes and problems. One way to think of this is to see the root cause as the main problem and the smaller causes as the symptoms to the problem. You will see yourself as the doctor that needs to fix the illness within the company. When you give the company their prescription, the symptoms will decrease and become eliminated. You can only truly eliminate the problem and help the company continuously improve if you focus on the root cause.

13 Key Practices

Even though many people are just learning about Lean Six

Sigma, the profession has been around for decades. In fact, global research of Lean Six Sigma has been a part of studies for over 20 years. The goal was to identify the best lessons and practices of Lean Six Sigma and how they can improve businesses. Throughout the study, it was found that these general practices helped improve every business. It is the 13 key practices that every Lean Six Sigma professional should follow to make a business successful and more effective.

1. Lean Six Sigma teams need to focus on the whole system. While you will break the system down into multiple parts, it is important that you do not forget about the bigger picture.
2. You need to make sure the time is right. Think back to the recycling and waste management case study and notice how much time employees wasted during various moments throughout their day. This is not to say that employees mean to waste time, it is a common problem in nearly every company. Wasting time disrupts the flow and the goal is to create an efficient flow for everyone involved.
3. Everyone needs to become involved in Lean Six Sigma practices. If you are helping a large company, you won't have everyone on your team. However, you should still include them when it comes to the analysis and improvements. You always have to focus on giving everyone the best efforts in communication between themselves and the Lean Six Sigma team.
4. The management team has to be excited about the prospects of Lean Six Sigma. If you can't get the management team of the company on board, the whole system can fail.
5. You always need to make sure you commit the top people of the company. Other employees will

become more invested with the Lean Six Sigma practices if they see their supervisors and leaders committed to the process.

6. Don't forget about hands-on training. Even if some employees have been trained in their job for years or decades, it is always a good idea to give every employee effective hands-on training that will help improve the system. Furthermore, it is always a good idea for all Lean Six Sigma professionals to receive their own training with the company as this can help you understand the company, its processes, and policies better.

7. You want to keep your improvements realistic. It is important to realize that improvements will take time and the team will bring themselves back to the drawing board often to come up with different practices as flaws can unfold during the improvement process as well. You want to take everything into consideration when you are focusing on improvements. For example, if you are looking at the flow of the organization, consider the job descriptions for each individual or area. You also want to consider if the employee is part-time or full-time. It also helps to consider their strengths and weaknesses as some employees can help improve the flow in one area better than another area.

8. Make sure all the measurements and statistics used are understood by everyone. There are no complicated statistics when it comes to Lean Six Sigma, but if someone doesn't understand how you came up with the data, you need to take the time to explain it to them.

9. Ensure that you take time to describe the Lean Six Sigma team and what everyone's responsibilities are from the beginning. Everyone should have a clear

idea of what their roles are and how to achieve the main goal.

10. Realize that the smaller roles and practices you incorporate into the organization will grow into larger practices. As long as the improvement efforts are working, the company will continue to grow and will always find a way to improve.

11. You need to realize that you will not always be a part of the company. Eventually, most Lean Six Sigma employees become a more silent partner. They help the team during the process and play a hand in the improvement, but most professionals back away and allow the company to thrive. Of course, some companies will hire a Lean Six Sigma professional full-time and this means that you could be a visual part of that team for years.

12. Always think about what a business can save when you are in the analyzing and improvement process. For most organizations, one of the goals is to save as much money as possible. This might be in a certain area or within the whole organization. Even if the team focuses on a certain area, it is important that you consider the whole organization as everything pieces together at the end.

13. If the company wants to look at various projects, you want to first choose a project that will build the business' credibility at a rapid rate. Once this project is in the improvement phase, you can start to look at the other projects.

Continuing Statistics and Analysis

Lean Six Sigma professionals help you set up a process and database that allows you to control your statistics and analysis for years to come. While you will find yourself changing software over time, you will continue to control your methods to make sure that your company is always on the right track. Here are a few tips to help you focus on your continuous progress, especially when it comes to collecting data and analysis.

Always Be Consistent

Once you put a plan into motion, you want to make sure you are following that plan throughout the project. There are always going to be days and moments where you don't have the energy to put forth the effort because you aren't feeling well, you didn't get enough sleep the night before, or you aren't completely interested in the task you are working on. When you feel a bit sluggish, you will find yourself becoming inconsistent. It is important that you do everything you can to make sure you remain consistent, even when you don't have as much energy or interest as you normally do. Consistency will make a project successful while inconsistency will break a project down.

Always Acknowledge the Results

You will not always like the results you receive for a project. There are times when even the most experienced Lean Six Sigma professionals are wrong about the hypothesis. It is important that you do not take these moments personally. You want to acknowledge the results as this will help the team develop the best plan for their improvement. Even if you need to go back to the drawing board, knowing that your hypothesis will not work out brings you one step closer to helping the company succeed.

Manage the Resistance to Change

Change is continuous when you are always working on improving the company. There are a lot of people who aren't comfortable with change, which can make Lean Six Sigma practices difficult for them to digest. As a Lean Six Sigma professional, it is up to you to help them through the changes and allow them to become more comfortable with changes. You will focus on various strategies to help people overcome their resistance to change.

- **Set challenging, yet achievable goals.** Throughout the Lean Six Sigma process, you want to make sure that the goals the Lean Six Sigma team and the company sets are challenging, yet achievable. You want to make sure that all the employees understand the goals and the steps that it will take to achieve the goals. Always be clear when you are giving the employees guidance as this will help them become more passionate about the goals. It is always helpful to set up one-on-one meetings to help the employees find their initiative and notice their individual progress. You also want to remember to set up milestones that are celebrated once its reached.
- **Persuasion is key.** To engage some employees, you will need to become an energized and persuasive leader. You will need to be focused on the opportunities and show the employees what the added work and changes in their routine can do for them and the company. It is always helpful to share experiences of other companies to get people motivated to try the same or similar process. When you are using persuasion, you need to focus on the positive change that will occur. While there might be some negative steps along the way, as long as you continue to focus on the positive changes, you will engage people to work hard and they will become determined to succeed.
- **Always remain supportive.** Along with focusing on the positive, you want to remain supportive,

especially when people are facing change. Employees are more likely to become confident that the change will help the company grow when they have a supportive leader who wants to help them through the process. Even when you find yourself becoming more of a silent partner, you want to continue to make yourself available to remain supportive when the company contacts you with any questions and concerns about their continuous improvement process. While these questions and concerns will start to diminish over time, they might contact you often at the beginning. It is important to show patience, even when you are focusing on a different company and their concerns and growth.

Resolve Any Conflicts Effectively and Quickly

One of the quickest ways to break down communication or a plan of action is by allowing conflicts to become a part of the process. There might be a time when you are trying to help a company succeed and some of the employees do not get along with each other. No matter what one employee says, the other employee will always criticize the idea. This can bring more challenges into the mix, especially for the Lean Six Sigma team. In fact, if the employees are constantly arguing, you can find the plan at a standstill as the other employees will not have motivation to work on the process. Furthermore, conflicts can waste valuable time in the process.

Some conflicts will be easy to solve, while others will be more challenging. When it comes to challenging situations, you will want to focus on the seven methods of the carefronting strategy. These steps will help you dissolve any conflict that stands in the way of a successful plan.

1. **Listen and understand.** Always take time to understand the person's point of view. The quickest way people become angry is when they feel like people are not listening and understanding what they are saying. If an employee brings up their opinion, it is important to them and they deserve to be heard. Their opinions tend to follow their own values and beliefs, which is something people are passionate about. Take your time when talking to them and realize that anyone can take the comments you make personally. If you have to pause the planning or any part of the process to discuss the emotions of employees to bring stronger support and open up the lines of communication in the workplace, then you need to be prepared to do so.

2. **Accepting and owning anger.** Once you start to notice someone becoming angry, you want to start focusing on getting the individual to release their anger. Anger is one of the driving forces when it comes to conflict and the sooner you can end the conflict, the easier the rest of the process will go. One of the most useful tools is to turn the anger into a positive emotion. You want them to accept and own their anger, which can also be a challenge. However, by focusing on the positives and remaining calm, you can help the individual control their anger and minimize any following conflict.

3. **Inviting change without demanding it.** One of the quickest ways to bring conflict into the workplace is by demanding the employees to change. Telling employees they need to change their behavior, attitudes, policies, or procedures—even when they request your assistance—is not an easy task. In fact, many Lean Six Sigma professionals feel the way they handle discussing the results of their analysis and the

process of improvement is the toughest part of their job because emotions can start to run high. People can quickly take offense to the Lean Six Sigma professional telling them that they are wasting time on their job. The way you talk to the employees about the changes can set the tone for the rest of the process. You need to talk to everyone about the change in a way that they can all agree it will help the company thrive.

4. **Becoming open to trust.** Trust is important when it comes to the workplace. It is a breakdown in communication and trust that typically causes most conflicts. You want to help the employees understand that each opinion is valued, especially when it comes to the individual. Once there is mutual trust, the team can focus on coming together to an agreement to end the conflict.

5. **Don't focus on blaming.** It is easy for people to place blame on each other. However, it is important that people move beyond the blame game and focus on reaching a compromise and moving forward. When people forget about who is to blame, everyone can come to an agreement.

6. **Accept responsibility.** When it comes to conflict, people need to take responsibility for their actions and what they say. The moment employees take responsibility, they begin to release the emotions that they hold onto during the conflict. They start to realize that no one, including themselves, is perfect and in turn, they become more open to compromising.

7. **Remain caring and encouraging throughout the compromising.** The only way conflict management is successful is when you keep the environment encouraging and caring. People need to feel like

someone is listening to them, understands them, and truly cares about what they are feeling and thinking. This is essential during compromising as there are a lot of points of view and people can easily allow their emotions to control them.

AFTERWORD

You are now completely prepared to take the next step towards your career and develop your Lean Six Sigma plan to help a company thrive. You are also ready to go a bit further into your career and establish your own business. The tools you have within the contents of this book provide you with valuable information that will lead you down the path of success.

Throughout this book, you learned everything that you need when taking your first step into your Lean Six Sigma career. You received various career opportunities, most of which continue to grow as the profession is growing. More people are becoming aware of the benefits of Lean Six Sigma, which brings them to look into receiving the certification or becoming a Six Sigma Champion or contacting someone like you.

As a growing profession, Lean Six Sigma is becoming a popular method in helping businesses find their weaknesses and turning them into strengths. For example, colleges are looking at ways to improve their student experiences by sending surveys and listening to the results. The healthcare field is doing what they can to save costs and improve patient experiences by changing their policies and procedures, such as the Toledo Medical Center did when they optimized their kidney transplant cycle time to 180 days or less. The government is also becoming involved in Lean Six Sigma practices, as you read when it came to their efforts in increasing production at a recycling facility.

It seems that everyone is starting to get on board with Lean Six Sigma practices as professors at universities are implementing these practices into their classrooms. Not only to help manage their classrooms, but to teach their students these practices. This is exactly what Dr. Evans and her students at Rose-Hulman Institute of Technology focus on every semester when it comes to monitoring the waste occurring at their college.

Startup businesses are also incorporating Lean Six Sigma practices into their business from the beginning. They want to know what they need to do to make their business thrive without struggling for years or possibly closing their doors two years down the road. They want to learn how to monitor their progress and catch their problems early, so they can actively find ways to solve any issues that come up.

One of the biggest reasons why Lean Six Sigma practices are becoming popular in nearly every type of business is because of the way it handles conflict and works to solve problems. The Lean Six Sigma teams understand the importance of communication and how it can quickly break down when employees get into their everyday groove and don't think about connecting with other employees. A lack in communication happens naturally when executives of a company know their employees are completing their tasks but are unaware of the process or procedures happening. Communication also breaks down with a small number of staff and everyone working hard to build up the company. People become so focused on their work that they don't think about sharing their processes.

When a company experiences a lack of communication, several problems will arise from this. The biggest problem of them all is no one is going to notice the issue until it starts to boil over, and employees are in conflict or frustrated because they feel that they are lacking support, or no one is informing them of any changes. Lean Six Sigma professionals understand the importance of catching these problems and eliminating them through a process. They understand the challenges employees face when they are confronted with the problems and find the root cause and its symptoms. However, through the efforts of Lean Six Sigma, businesses can overcome these challenges and become comfortable with change. They will become determined to see the business grow and will want to continue monitoring the progress.

No matter where you are in your Lean Six Sigma experience, there is information in this book that helped you. For example, you may have learned the best way to handle conflict when it arises in meetings. You may have learned how to help employees who resist change. You understand how Lean Six Sigma can help nearly every business and are prepared to take the next stage in your career.

Some of the key takeaways from this book that you should bring with you into every Six Sigma job are:

- Never accept defects. It doesn't matter if it is a lack of communication or broken materials. If you accept defects in the process, everything will slowly start to break down and you will need to start over. Make sure that you do everything thoroughly the first time.

- Consistency is key. When you are consistent about your methods, you will help employees become consistent with the methods as well. This consistency will continue once you become more of a silent partner because the methods are now a habit.
- Always look for the root cause of the problem. You will find many causes, but most of them are symptoms. When you find the root cause, you can eliminate the symptoms and focus on fixing the problem.
- Always start meetings by talking to the employees about Lean Six Sigma and what you can do to benefit their company. When employees note what you can do for them and their company, they will become passionate about the plan and work to reach the goals.
- Patience is required as a Lean Six Sigma professional. This is something that is often overlooked but is important throughout the process. Remember, the employees are not trained in Six Sigma and you will need to start with the basics so they can understand the process.
- Always believe in yourself and your efforts. When you believe in yourself, the employees of the company you are helping will believe in themselves and the project will become successful.

REFERENCES

9 benefits of recycling | Friends of the Earth. (2018). Retrieved 29 October 2019, from https://friendsoftheearth.uk/natural-resources/9-benefits-recycling

10 top Lean Six Sigma tips worth remembering. Retrieved 24 October 2019, from https://www.bsigroup.com/en-GB/blog/Lean-Six-Sigma-Blog/10-top-Lean-Six-Sigma-tips-worth-remembering/

A Look at Six Sigma's Increasing Role in Improving Healthcare. (2019). Retrieved 21 October 2019, from https://www.villanovau.com/resources/six-sigma/six-sigma-improving-healthcare/

Aminu. Six Sigma in Healthcare: Concept, Benefits and Examples - Dr Aminu. Retrieved 22 October 2019, from https://draminu.com/six-sigma-in-healthcare/

Business Process Manager Overview and Job Outlook. (2019). Retrieved 21 October 2019, from https://www.villanovau.com/resources/bpm/business-process-manager/

Carver, M. Six Steps to Effectively Plan for Lean Six Sigma

Efforts. Retrieved 27 October 2019, from https://www.isixsigma.com/new-to-six-sigma/deployment/six-steps-to-effectively-plan-for-lean-six-sigma-efforts/

Clark, S., Feldman, J., & Johnson, K. 10 Challenges to Overcome when Deploying Lean Six Sigma in Pharmaceutical Sales and Marketing - iSixSigma. Retrieved 27 October 2019, from https://www.isixsigma.com/industries/healthcare/10-challenges-overcome-when-deploying-lean-six-sigma-pharmaceutical-sales-and-marketing/

College Students Use Six Sigma to Solve Food Waste, Recycling Challenges. (2017). Retrieved 27 October 2019, from https://www.sixsigmadaily.com/college-students-six-sigma-food-waste-recycling/

Dillard, J. The Data Analysis Process: 5 Steps To Better Decision Making. Retrieved 30 October 2019, from https://www.bigskyassociates.com/blog/bid/372186/The-Data-Analysis-Process-5-Steps-To-Better-Decision-Making

Districts Turn to Process Improvement for Better Student Outcomes. (2018). Retrieved 30 October 2019, from https://www.sixsigmadaily.com/schools-process-improvement-student-outcomes/

Franchetti, M. (2015). Lean Six Sigma for Managers and Engineers with Applied Case Studies (1st ed., p. Kindle Edition.).

Gallo, A. (2015). A Refresher on Regression Analysis. Retrieved 28 October 2019, from https://hbr.org/2015/11/a-refresher-on-regression-analysis

Goodman, P. (2019). Top 10 Reasons Why You Should Recycle Your Waste. Retrieved 29 October 2019, from https://owlcation.com/stem/10-Reasons-Why-You-Should-Recycle-Your-Waste

Graves, A. (2014). Careers That Benefit From Six Sigma Training. Retrieved 20 October 2019, from https://www.sixsigmadaily.com/jobs-in-six-sigma-which-careers-utilize-six-sigma-training/

Hessing, T. Types of Data | What you need to know for Six Sigma certification. Retrieved 27 October 2019, from https://sixsigmastudyguide.com/types-of-data/

Leblanc, R. (2019). How Recycling Creates New Jobs. Retrieved 29 October 2019, from https://www.thebalancesmb.com/recycling-and-new-job-creation-2878003

Maroda, S. (2019). Director of Six Sigma Operational Excellence. Retrieved 24 October 2019, from https://www.cybercoders.com/director-of-six-sigma-operational-excellence-job-498245

Maroda, S. (2019). Director of Six Sigma Operational Excellence. Retrieved 24 October 2019, from https://www.cybercoders.com/director-of-six-sigma-operational-excellence-job-498245

Personal Goal Setting: – Planning to Live Your Life Your Way. Retrieved 27 October 2019, from https://www.mindtools.com/page6.html

Schembri, J. (2012). What Does a Project Manager Do?. Retrieved 22 October 2019, from https://www.sixsigmadaily.com/project-manager-job-description/

Schembri, J. (2015). Six Sigma Job Description. Retrieved 20 October 2019, from https://www.sixsigmadaily.com/six-sigma-job-description/

Sherman, P. (2014). 10 Reasons Organizations Do Not Use

Lean Six Sigma. Retrieved 27 October 2019, from https://www.qualitymag.com/articles/91986-reasons-organizations-do-not-use-lean-six-sigma

Six Sigma in Education. (2018). Retrieved 29 October 2019, from https://www.villanovau.com/resources/six-sigma/in-education/

Six Sigma Students Uncover Ways to Cut Waste on Campus. (2015). Retrieved 29 October 2019, from https://www.rose-hulman.edu/news/2015/six-sigma-students-uncover-ways-to-cut-waste-on-campus.html

Surendran, A. Qualitative Data- Definition, Types, Analysis and Examples. Retrieved 27 October 2019, from https://www.questionpro.com/blog/qualitative-data/

The 5 most common Six Sigma mistakes and how to avoid them. Retrieved 27 October 2019, from https://www.lean-sixsigmagroup.co.uk/5-six-sigma-implementation-mistakes/

Value Stream Analysis| Lean Six Sigma Value Stream | Quality America. Retrieved 28 October 2019, from https://qualityamerica.com/LSS-Knowledge-Center/lean-sixsigma/value_stream_analysis.php

What Business Analysts will learn from Lean Six Sigma Certification Course | Lean Six Sigma, Six Sigma Certification. Retrieved 20 October 2019, from http://www.sixsigmacertificationcourse.com/what-business-analysts-will-learn-from-lean-six-sigma-certification-course/

CPSIA information can be obtained
at www.ICGtesting.com
Printed in the USA
BVHW052250090223
658263BV00007B/306